D1267967

Engineer Your Own Success

Engineer Your Own Success

7 Key Elements to Creating an Extraordinary Engineering Career

Anthony Fasano, PE

PRO▪COMM
IEEE Professional
Communication
Society

IEEE PRESS

WILEY

Library of Congress Cataloging-in-Publication Data is available.

ISBN: 978-1-118-65964-9

10 9 8 7 6 5 4 3 2 1

Contents

A Note From The Series Editor

The IEEE Professional Communication Society (PCS, also known as ProComm), with Wiley-IEEE Press, continues its book series titled *Professional Engineering Communication* with Anthony Fasano's book *Engineer Your Own Success: 7 Key Elements to Creating an Extraordinary Engineering Career.* This book is like a breath of fresh air that will change the very way that you think about words like *success, organization, networking, mentoring,* and *professional communication.*

From my perspective, as someone who has seen hundreds and hundreds of graduating engineers come and go, I will be recommending this book far and wide. For those who are already well-established in their careers, this book will give you the push needed to make powerful and positive changes in your own work habits and future goals. With an encouraging tone, Fasano steps everyone through the processes that will help people study for professional exams, find a mentor, be a mentor, network, improve communication strategies, and march assuredly toward a self-defined arena where work is rewarding, both financially and personally. As I edited this book, I changed many of my own habits, my too-familiar ways of thinking, and mentoring habits. And old dog can learn new tricks!

From a larger perspective, this book is a welcome addition to the *Professional Engineering Communication* (PEC) book series, which has a mandate to explore the areas of communication practices and application as applied to the engineering, technical, and scientific professions. Including the realms of business, governmental agencies, academia, and other areas, this series will develop perspectives about the state of communication issues and potential solutions when at all possible.

All of the books in the fast-growing PEC series keep a steady eye on the applicable while acknowledging the contributions that analysis, research, and theory can provide to these efforts. Active synthesis between on-site realities and research will come together in the pages of this book and others in the series, past and future. There is a strong commitment from PCS, IEEE, and Wiley to produce a set of information and resources that can be carried directly into engineering firms, technology organizations, and academia alike.

For the series, we work with this philosophy: at the core of engineering, science, and technical work is problem solving and discovery. These tasks require, at all levels, talented and agile communication practices. We need to effectively gather, vet, analyze, synthesize, control, and produce communication pieces in order for any meaningful work to get done. It is unfortunate that many technical professionals have been led to believe that they are not effective communicators, for this only fosters a culture that relegates professional communication practices as somehow secondary to other work. Indeed, I have found that many engineers and scientists are fantastic communicators because they are passionate about their work and their ideas. This series, planted firmly in the technical fields, aims to demystify communication strategies so that engineering, scientific, and technical advancements can happen more smoothly and with more predictable and positive results.

Traci Nathans-Kelly, Ph.D.

Acknowledgments

I would like to acknowledge the many people in my life who have supported me, not only in creating this book but in my journey to inspire engineers, as well as those who have mentored and guided me through my career and my life.

First and foremost, I'd like to thank my wife, Jill, who has supported me in every aspect of my life and career. She has believed in me regardless of how big I dream, even when others have doubted me. To my three children, Brianna, AJ, and Penelope, who are a constant source of inspiration, I'd like each of you to know that you bring endless joy to my life.

Thank you to my parents, Tony and Rita, who have given me constant support and encouragement throughout my life and who have always reminded me that I could do anything I want if I put my mind to it. Also, to the rest of my large family including my brothers, Christopher and Michael; mother-in-law, Michele; father-in-law, Brian; sisters-in-law, Danielle and Kim; aunts; uncles; cousins; and grandparents, thank you for being in my life. Many of you were there to provide input on this book and help me on my journey whenever I needed it.

I would also like to thank Richard Maser and all of the Maser Consulting staff for their support and friendship during my 10 years with the organization. It was such a pleasure to work for a company that cares so much about the development of its employees. My time spent there had a huge influence on this book and on my career.

I would like to thank all of those individuals who have inspired me and helped me to grow including professors, colleagues, fellow engineering society members, fellow coaches, clients, and friends.

I would like to thank the talented team that helped me to take my dream of being an author and make it a reality through the words you are reading in this book. The team included my wonderful coach, Kathleen Gage; the marketing team of Lauren McMullen, Doug Yuen, and Melanie Borowczyk; and those who read over the manuscript and provided me their valuable input, including Marlene Oulton, Patricia Moran, Dr. James Yarmus, Bernie Berson, and Merlin Kirschenman.

I would also like to thank Carl Selinger for referring me to Wiley-IEEE Press and Jill Bagley and Mary Hatcher for making the process of writing this book much easier than I thought it would be. Last, I would like to thank my editor, Traci-Nathans Kelly, who immediately understood the message I was trying to convey to engineers through this book and helped me to do it in a powerful way that I believe will impact engineers around the world for years to come.

Foreword

Anthony Fasano, PE, is a professional engineer and former employee of Maser Consulting P.A. Today, after following his true calling, he has become a recognized writer and inspirational speaker. In his book, *Engineer Your Own Success: 7 Key Elements to Creating an Extraordinary Engineering Career*, he breaks down the fundamentals of a proactive and commonsense approach toward all of the things they don't teach you when obtaining an engineering degree. Aside from the technical skills you learn in school, there are certain actions that go hand in hand with making a conscious effort in growing your career.

In 2010, during his tenure with Maser Consulting, Fasano became a certified professional coach and developed and implemented a program within our firm called Energy Leadership. He worked with over a dozen key employees, focusing on transitioning their roles into leadership positions. This program successfully demonstrated how to take your career into your own hands through awareness of your professional surroundings, use it to your advantage to become a leader, and motivate those around you to reach their full potential.

Becoming successful almost never just happens without some strategic effort. Anthony's book is something to reference throughout your career. He is a testament to his own positive teachings as he has shown that you can change careers and be successful by following some very basic elements that can be applied across the board.

Richard M. Maser, PE, PP, CME
CEO and President
Maser Consulting P.A.

Preface

The last few years have been an amazing journey for me. I have gone from being the youngest associate at a well-regarded engineering consulting firm to my dream job of inspiring engineers worldwide in their career development efforts. A series of events led me to take this leap of faith to chase my ultimate dream of helping engineers, and I want to share my story with you.

Upon graduating from Lafayette College in 2000, I started working as a civil engineer for a reputable firm in New York. As a determined professional, I made a conscious decision to advance my career as rapidly as possible, which I did, earning several licenses, certifications, and awards along the way.

However, in the process, there were certain aspects of the engineering industry that bothered me. First, while many engineering companies support their employees' career development, many don't, and consequently, too many younger engineers feel alone or lost early on in their careers. I believe that part of the reason this happens is that the nontechnical skills engineers need to advance aren't taught or emphasized at most schools. Second, I noticed that the nature of the industry, specifically the long hours and tight deadlines, affected many engineers on a personal level with the countless stressful hours proving detrimental to their family lives. Finally, I noticed that most engineers focus so much on their jobs that they forget about their careers, which become stagnant and inhibit company and personal growth. I was determined to change these things in a big way, but first, like any good engineer, I had to figure out how to solve these problems.

About two years ago, the beginning of the solution came to me through a request from my employer at the time. The company noticed my rapid advancement and asked me to deliver a seminar on career development to engineers throughout the firm. I delivered these presentations as requested, and not only did I thoroughly enjoy doing so, but the engineers who attended started taking positive steps in their careers.

It was at this time that I realized I had an ability to help professionals move forward, so I enrolled at a top coaching school where I went through an intensive training program and became a certified professional coach. For those of you not familiar with executive coaching, it is a process in which a trained professional helps bring out the very best in another person or group of people, whether the trainees are individuals, businesses, teams, or organizations. Coaching is an action-oriented process with a focus on the client's current situation as well as short- and long-term goals. My own short-term goal at that time was to launch a coaching company that engineering firms would retain to provide coaching services with an emphasis on career and leadership development. This would enable companies to provide career development support to their engineers, which would help them stay engaged, productive, and efficient, all of which would benefit both the companies and their employees.

Providing services to engineering companies would be half the battle; I knew that I would also have to reach out directly to engineers. To do this, I took a two-tiered approach: speaking and writing. I had always enjoyed public speaking, but I knew that to be a highly effective speaker and reach large numbers of engineers at one time, I would have to improve. I studied a multitude of books on the subject and fine-tuned my skills by joining a public speaking group. Over the past five years, I have been presenting seminars focused on inspiring engineers in their career advancement and providing them with information on topics that I believe are critical to success: setting career goals, mentoring, credentials, networking, organization, communication, and leadership.

Last but not least, I write—and I write a lot. I maintain a blog and associated podcast on my website, www.engineeringcareercoach.com, that is dedicated to career advancement for engineers. My blog features articles written by me and by other professionals in the engineering industry. In addition to my blog, I provide a free service to engineers and other professionals worldwide called *A Daily Boost from Your Professional Partner.* These *Daily Boosts* are written by me posing as "your professional partner," a mentor guiding you through your career. They are short personalized e-mail messages that go out each morning, Monday through Friday. Each of the Daily Boosts contains a career advancement tip and/or inspirational message with the aim of helping engineers and other professionals to focus on their career each and every day. You will see some of these inspirational notes at the end of each chapter and you can sign up for them yourself at www.dailyboosts.com.

Over the past few years, I have achieved all of my goals. In July 2010, I left my job as a civil engineering associate at a consulting engineering firm and started my own company, Powerful Purpose Associates, which led me to founding the Institute for Engineering Career Development (IECD). The IECD is the only community that provides career coaching and training for engineers while also giving them access to a private community of like-minded engineers to help them achieve their goals. Since launching my company, I have spoken to more than 40 engineering societies and organizations throughout the United States, including national and international conferences, and the IECD has helped countless engineers identify and achieve their career goals while creating life-long friendships in the process.

Morphing from being an engineer to inspiring engineers has been an exciting journey—my career now is a dream come true. I have been able to meet and connect with engineers around the world and help them to create successful, enjoyable, well-balanced careers.

It is now your turn. In this book, I give you all of the information and steps that I have taken to get to where I am today so that you can use them to build your dream career. This information includes strategies for improving the core skills that engineers sometimes struggle with including goal setting, finding a good mentor, networking, delegating, and public speaking. Career advancement is easier and more fun than you think, so buckle up for a wonderful ride!

To your success,
Anthony Fasano, PE
The Engineering Career Coach

Introduction: Use This Book Strategically

As an engineer myself, I know that it can be extremely difficult to advance in the field of engineering even though we come out of school with such an extensive technical background.

Especially early on in our careers, we know little but numbers and equations. However, in order for us to advance upward on the corporate ladder, we need to increase our value by focusing on developing our core skills in addition to our technical expertise. Core skills are those that help us to communicate, to build relationships, and, most importantly, to develop our ability to be great leaders.

This book defines the seven elements that I believe are critical to your success. They cannot be learned in a textbook or a laboratory. I learned them myself through years of practicing them and studying other experts in these fields.

Part I of the book provides strategies for finding an engineering job, which you may need to utilize at different times throughout your career. Whether you are a senior engineering student, an unemployed engineer, or a seasoned veteran looking to make a move, this part of the book will be helpful to you. It includes recommendations for résumé preparation and the actual interview process, many of which have been given to me by engineering hiring managers and human resource professionals. If you are happy in your current engineering job and are simply looking to advance, skip to Part II.

Engineer Your Own Success: 7 Key Elements to Creating an Extraordinary Engineering Career,
First Edition. Anthony Fasano.
© 2015 The Institute of Electrical and Electronics Engineers, Inc. Published 2015 by John Wiley & Sons, Inc.

The seven key elements to creating an extraordinary engineering career include career goals, credentials, mentoring, communication, networking, organization, and leadership. Part II of this book is devoted to defining each of these key elements and associated core skills, while Chapter 11 provides the tools and templates to take action and utilize this information. The tools can also be downloaded at engineeringcareercoach.com/EYOS. While the seven elements are laid out in a fairly logical order, they certainly do not have to be followed in that order. In fact, you may develop many of them at the same time.

Chapter 12 of the book contains short success stories from engineers that have implemented some of these tools after reading the first edition of this book. I am thankful that they have taken the time to share them with me. Finally, Chapter 13 of the book contains some of the most popular blog posts from the engineeringcareercoach.com blog that I have been told by engineers have helped them tremendously in their career.

It is my sincere desire that you take the steps and strategies in this book and apply them to your career so that you can be as successful as you want to be.

Part I

YOUR GUIDE TO ENGINEERING A SUCCESSFUL JOB SEARCH

1

Building a Winning Résumé

When it comes to getting a job or client, congruent value is aligning the employer's need with your value add.
　　　　—Richie Norton, Resumes Are Dead and What to Do About It [1]

1.1　Building a Winning Résumé (Online and Offline)

People often talk about a "winning résumé." Is there really a difference between a typical résumé and one that "wins" over the mind of the prospective employer?

My definition of a winning résumé is very simple: it helps you land the job that you are applying for. To be more specific, a winning résumé is a document that helps to secure an interview. This step alone is a huge challenge, especially in a down economy or in a competitive market.

A winning résumé does not look a certain way, nor is it a certain length; it is a résumé that showcases your qualifications and ultimately results in an interview. No matter how it looks or how long it is, if you send it in and do not get an interview, it is a losing résumé—end of story.

In the following pages, I will give you a blueprint for building a winning résumé for any job that you apply for. While I will offer some guidelines as to the appearance and length

Engineer Your Own Success: 7 Key Elements to Creating an Extraordinary Engineering Career,
First Edition. Anthony Fasano.
© 2015 The Institute of Electrical and Electronics Engineers, Inc. Published 2015 by John Wiley & Sons, Inc.

of the résumé, your overall focus on the target person or company who is doing the hiring is the most important aspect of this résumé preparation process.

> "Find out what you like doing best and get someone to pay you for doing it."
> — Katherine Whitehorn   [2]

1.2   There Is One Key Factor to a Great Résumé

As part of my job as an engineering career coach, I have had the opportunity to speak to many engineering hiring managers and industry recruiters, and I always ask them to explain to me exactly what they are looking for in a résumé. You may be surprised at what they have told me.

While the content of your résumé and your experience are critical to being hired by an engineering company, recruiters and hiring managers say that it is the visual presentation that will actually get you in the door for that all important first interview.

> While the content of your résumé and your experience are critical to being hired by an engineering company, recruiters and hiring managers say that it is the visual presentation of your résumé that will actually get their attention.

There are hundreds of engineers applying for the same job these days. Recruiters and hiring managers don't have the time to look through hundreds of résumés, so instead, they scan them first. They spend about 10 seconds per résumé, identifying which ones look good enough for them to come back to for a more detailed review at a later date.

The résumés that typically make it to the second round of reviews are those that are neat and easy to read and contain some key points about the individual. They simply stand out from the rest.

There is no right or wrong way to create a résumé; the key is to make it presentable and give the reviewer some points that will create interest. When creating your résumé, keep in mind that *someone may look at it for a mere 10 seconds and decide whether or not you will be interviewed.*

1.3   The Importance of Customizing Your Résumé

Before we get into the actual résumé preparation, I want to convey the importance of customizing your résumé for a specific desired job. Here's a true story. During one of my *Engineer Your Own Success* seminars, an engineering executive in the audience shared some very powerful advice with the attendees. This executive worked for a large engineering company, and one of his responsibilities was to review résumés of prospective candidates and decide which ones his company should interview.

He told us that, in one instance, while looking through a pile of résumés, a candidate mentioned one of the projects that the executive's company was working on at the time. Let's say, e.g., the executive worked for XYZ Company. The applicant's résumé said, "I am currently managing the John Smith Memorial Hospital project which is very similar in nature to XYZ's Bill Taylor project in that it has a budget of $X,XXX,XXX and the clients goal is to ABC." The executive told us that as soon as he saw his company's name on the résumé, he decided to interview this candidate. Whether or not this candidate got the job is irrelevant because getting the interview can mean winning more than half the battle. The larger point is this: the applicant made a brilliant move by linking personal work projects to the current projects of the hiring company. *Nothing impresses a company more than itself, so use that to your advantage.*

Experienced recruiters have told me that one engineering position could have hundreds of applicants, so following this executive's advice could mean the difference between being considered for the interview or getting passed over. To stand out, you need to customize each résumé to match the requirements of the company where you are applying. I believe you will find it is well worth the extra time.

> To stand out, you need to customize each résumé to match the requirements and stated goals of the prospective employer.

This should be fairly simple to do if the position you are applying for has a complete job description. Simply use the same words from the job description on your résumé; those keywords, already in the job advertisement, tell you how to word your own application documents. Look for repeated wording like "leadership" or "technical competency" or "excellent communication skills." Then, repeat those keywords or phrases, as applicable, in your own résumé and letter. If the job ad asks for someone with industry contacts, explain on your résumé or in your letter how you have a large network in the industry (if it's true).

I have always told engineers that if you apply for 10 different jobs, you should have 10 different résumés. If you don't, you are not putting yourself in the best position to land the jobs.

1.4 There Is a Formula to Building a Winning Résumé

Here are seven tips for creating a winning résumé or modifying an existing one:

1. *Include your contact information in the header section of the document.* It should appear on all pages and should not cut into valuable résumé space. Ensure that your information is neatly listed and avoid offensive e-mail addresses (i.e., lazy@ad.com). Hint: websites like Engineering.com allow you to get a free e-mail address with their URL (@engineering.com).
2. *Left justify the text.* Doing so ensures that the text is neatly aligned. When you are counting on someone processing the information on your résumé in a short

period of time, neatness counts. It must be neat and easy to read. Invisible tables can help you align segments of text.

3. *Ensure that the layout of the résumé is consistent.* For example, if the date range for a job is in one place, be sure that it is in the same relative location for all positions you have held over the years. Using tables to create a consistent look is a good technique, but remember to make the gridlines invisible.

4. *For professionals, place your most relevant work experience at the top of the listings.* Students and recent graduates should place the education section near the top of your résumé with experience (i.e., internships) immediately below. Please note that it is perfectly acceptable and sometimes encouraged to place non-engineering work experience on your résumé (see Section 1.6 for recommendations on effectively showing non-engineering experience).

5. *Use numbers where applicable.* For example, if you are currently managing a project with a $35 million budget, working on a senior design project with five other students, or you are currently managing a team of 12 project engineers, include those numbers. Numbers give reviewers tangible items that can provide them with an idea of the magnitude of projects you have worked on or managed.

6. *Bold important items that you want to stand out.* **This is important.** Remember, they are looking at it for only 10 seconds, so bolding text can really help catch a reviewer's eye. Be sure not to overuse this technique or the reviewer may ignore all of the bold points.

7. *Lose your personal career objective or make it count.* Recruiters and hiring managers have told me that everyone has the same objective (to get a job); therefore, they skip over any text about this aspect when doing their 10 second review. If you want to keep a career objective on your résumé, make sure it speaks specifically to the position you are applying for or highlights a unique skill or talent that sets you apart from others. Otherwise, eliminate it. I will discuss this in more detail in Section 1.5 as there are some situations where an objective is beneficial.

Remember, there is no right or wrong way to build a winning résumé. However, these guidelines have proven to be successful in my own experience as well as that of many other engineers that I have coached through this process.

At the end of Section 1.5, you can see a copy of my résumé. Please feel free to use it as inspiration when compiling your own. See also Chapter 11 in this book.

1.5 Determining the Proper Length of a Résumé

I can't tell you how many e-mails I get from engineers asking me how long their résumé should be. In an effort to answer this question effectively, I always ask hiring managers and recruiters this same question when I speak with them. Here is what they have told me and some ways that you can save space.

If you are an engineering student, your résumé should be no longer than one page. Keep it simple and follow the guidelines in the previous section. In the rare case that you had extensive work experience prior to or during college, more than one page may be acceptable. Be sure to include all of your extracurricular activities, especially those in engineering-related societies and clubs.

> If you are an engineering student, your résumé should be no longer than one page.

If you are an engineer in the first 5 to 10 years of your career, it is perfectly acceptable to have a résumé that is multiple pages in length (unless the job ad states otherwise). I would strongly suggest that you try to keep it to about two or three pages. Include as much detail as you can about projects that you have worked on or managed, including your responsibilities, project budgets, number of team members/consultants involved, timeframe, etc. Unlike a college student, your work experience should be at the top of the résumé before education.

For those engineers that have been working for over 10 years, you may have a résumé that is even longer, perhaps 5–10-pages long. Believe it or not, I have actually seen résumés that were longer than that. Similar to a younger engineer, the résumé should include details for all (or the most prominent) of the projects you have worked on or managed. You should also include a section on nonbillable activities that you are responsible for as well. Nonbillable activities that prospective employers might be interested in seeing include mentoring younger engineers, proposal preparation, billing, business development, and marketing activities, to name a few. (These longer résumés are typically referred to as CVs, which is short for the Latin "curriculum vitae.")

> Nonbillable activities that prospective employers might be interested in seeing include mentoring younger engineers, proposal preparation, billing, business development, and marketing activities, to name a few.

So how do you ensure that your résumé doesn't exceed the lengths described earlier in this section? Here are some tips for keeping your résumé at a reasonable length:

- *Remove the personal career objective or make it unique.* As I mentioned in Section 1.4, hiring managers have told me it is perfectly acceptable to remove the objective. They point out that everyone has the same objective—to get a job. Therefore, they don't spend much time, if any, reading it.

That being said, if you are applying for a position in a larger company where there are jobs with identification numbers or very specific titles, then it would be beneficial to have an objective that includes these identifiers, ensuring that the résumé gets to the right person.

If you use an objective, please make sure it is unique. Highlight something about yourself that would interest the company beyond the fact that you are looking for a position where you can build your career. Mention a skill that you have or situation that you have been in that separates you from others. It can also be beneficial to include the words "or related position" in case you don't end up being a good fit for the specific position you are interviewing for. This would alert the prospective employer to the fact that you are flexible and could be considered for other openings. As an example, the box below provides the objective on a résumé from one of the engineers I have worked with in this area; you can read his story entitled "Realizing a Dream of Becoming a Structural Engineer" in Chapter 12 of the book.

> Fast learning, motivated younger engineer, looking for an opportunity as a structural engineer after growing up in a structural engineering household with a father who ran his own engineering company.

- *You do not have to put the phrase "References available upon request" at the bottom of your résumé.* This is now a given fact; however, you should have a list of references prepared to give a prospective employer in an interview. Three is the usual expectation.

- *List only the projects and experiences on your résumé that are relevant to that position.* I will continue to repeat this advice, as you should have a different résumé for every job you are applying for. This will allow you to select only the projects that are applicable to the job you are trying to target. Not only will this cut down the length of your résumé, it will also present you as a better fit for that specific job.

- *You can decrease your margin size slightly, as long as the overall appearance is still neat and readable.* Make sure you justify the text on each page to keep it clean and easy to read. Avoid reducing the font to less than size 10 pt.

Anthony Joseph Fasano, P.E., LEED AP, ACC
123 Smith Drive
Smith, NJ 90210
Phone: 201-555-5555 E-mail: afasano@powerfulpurpose.com

SUMMARY

Anthony Fasano is a professional engineer who is now a well known engineering coach and consultant focusing in the areas of business growth and career development. He uses his engineering background coupled with his executive coaching skills and online marketing expertise to help engineering organizations sustain, grow, and expand in a powerful and positive way.

PROFESSIONAL EXPERIENCE

Powerful Purpose Associates, Ridgewood, NJ   **March 2009 – Present**
Managing Director/CEO
- Author of both the bestselling book Engineer Your Own Success: 7 Key Elements to Creating an Extraordinary Engineering Career and one of the top blogs on Engineering Career Development
- Utilize executive coaching skills to empower engineers to be both productive and efficient in advancing their careers, and to have a positive view of both their organizations and co-workers
- Created the Institute for Engineering Career Development and built it to almost 100 paying members in 4 months without doing any advertising or providing any PDH's
- Nationally recognized engineering speaker on the topics of career and leadership development

Maser Consulting P.A., Chestnut Ridge, NY   **July 2000 – July 2010**
Project Manager/ Civil/Site Engineer/Career Development Trainer
- Responsible for day-to-day operations of multiple projects from initial conception through final design
- Managed project teams and relationships and address concerns with county officials, county engineers, utility companies, historical societies and permitting agencies

FIGURE 1.1 Sample résumé of Anthony Fasano, part 1. This is a sample of the top of the first page of the résumé.

- *Have a colleague proofread it and ask that person what information really stands out and which information isn't interesting.* Such advice may help you to remove some extraneous details. It may even be more advantageous to have someone outside of the engineering industry to review it to give you the most unbiased advice.

A résumé does not have to be a certain length in order to be a winning résumé; however, your résumé should be an appropriate length for the amount of experience you have (Figs. 1.1, 1.2, and 1.3).

- Presented company-wide career development seminars increasing productivity and boosting morale

PROFESSIONAL LICENSES, MEMBERSHIPS AND AFFILIATIONS

Professional Engineering License (PE) in New York and New Jersey
LEED Accredited Professional (LEED AP)
New York State Society of Professional Engineers (NYSSPE)
American Society of Civil Engineers (ASCE)
Engineering EXPO Committee
Associate Certified Coach (ACC)
International Coach Federation (ICF)

SOCIETY ACTIVITIES

Acting President for NYSSPE Rockland County Chapter (4th term). Responsibilities include overseeing day-to-day operations including correspondence with State chapter and various industry leaders.

Treasurer and Secretary for NYSSPE Rockland County Chapter from 2004 to 2009. Responsibilities included managing finances for organization (bank account +/- $20,000).

Treasurer for the Engineering EXPO Committee from 2004 to 2009.

Founded ASCE YMG in 2004 and served as co-chairman for ASCE YMG from 2004 to 2006.

FIGURE 1.2 Sample résumé of Anthony Fasano, part 2. This is a sample of the bottom of first page of the résumé.

Anthony Joseph Fasano, P.E., LEED AP, ACC
123 Smith Drive
Smith, NJ 90210
Phone: 201-555-5555 E-mail: afasano@powerfulpurpose.com

EDUCATION

Institute for Professional Excellence in Coaching (IPEC)   **March 2009 – October 2009**
Certified Professional Coach
Energy Leadership Index Master Practitioner (ELI-MP)

Columbia University, New York, NY   **September 2000 - May 2002**
Master of Science in Construction & Engineering Management

Lafayette College, Easton, PA   **September 1996 –May 2000**
Bachelor of Science in Civil & Environmental Engineering
Vesalius College, Brussels, Belgium   Spring 1998

Don Bosco Preparatory High School, Ramsey, NJ   **September 1992 – May 1996**

RESEARCH

During Master's Degree program performed Independent Study regarding the impact of technology on engineering and construction. Study focused on internet construction management software.

AWARDS AND CERTIFICATIONS

Received 2010 Presidential Citation from New York State Society of Professional Engineers.

Received Forty under Forty award from Rockland County Economic Development Corporation in October 2009 for commitment to professional excellence and community involvement

FIGURE 1.3 Sample résumé of Anthony Fasano, part 3. This is a sample of the second page of the résumé.

1.6 Effectively Show Non-engineering Experience on Your Résumé

Engineers often ask me if they should list non-engineering jobs on their résumé when trying to land an engineering job. For example, what if you spent a few summers flipping burgers at Burger-o-Rama or were unemployed for so long that you took a job as a clerk at a supermarket? Should you really put these types of jobs on your résumé? In my opinion, the answer is *yes*, but they must be put on your résumé in a certain way, not simply listed under a section called Work Experience.

Here's how to strategize that work: create two different sections on your résumé. One can be titled "Engineering Work Experience" and another one titled "Non-engineering Work Experience." Using two sections to delineate your work categories allows you to identify for the potential employer-related work plus other works; doing so communicates a work ethic that might have been otherwise overlooked.

The other important aspect of putting unrelated work experience on your résumé is the way you write the job description. For the purposes of this example, let's say you are a recent college graduate trying to land an engineering job. Because internships were hard to come by, you spent the last two summers working at a fast-food restaurant. There, your job responsibilities included opening up the restaurant and setting it up for the day, cleaning the kitchen, taking and fulfilling customer orders at both the drive through and the register, and, yes, flipping burgers if the need arose or, to put it more eloquently, assisting in food preparation.

Here is an example of how not to show work experience.

Burger-o-Rama—Summer of 2010

I was unable to find an engineering summer internship and therefore worked as a waiter/cleaner at Burger-o-Rama where I performed the following tasks:

- Open up and set up
- Cleaning of kitchen
- Food preparation
- Fulfilling of customer orders

Here is a revised version that shows skills learned.

Burger-o-Rama—Summer of 2010

I worked as an assistant to the manager at a local Burger-o-Rama for the entire summer.

Overall, I believe that what I learned in this role will be very helpful to my success as an engineer.

(Continued)

Below is a list of my job responsibilities and what I learned from each one of them:

- *Open up and set up of restaurant.* I learned how important preparation is in a business and how it impacts the organization's bottom line.
- *Cleaning of kitchen/facility.* It is very important to take good care of your tools and equipment, and how failure to do so can have a devastating effect on a business and the products and/or services it produces.
- *Food preparation.* Proper food preparation made me realize how much responsibility I had in the growth of the business. If the food items aren't prepared properly, the customers won't be happy, and if customers aren't happy, the business will not succeed.
- *Fulfilling of customer orders.* This was my first experience dealing directly with customers, and I learned that above all else, the customer is your top priority and you must keep them happy. This really kept me on my toes.

When I first took this job, I thought it was going to be solely a source of income; however, it turned out to be an amazing learning experience and I am excited to implement what I have learned in the engineering world.

Now that example may be a bit lengthy, but I want you to really see the strategy. It isn't just about writing up a good job description; it's about knowing how to favorably list key elements you have learned from that position. If you read the job description earlier, the information is 100% true. In other words, if you have worked at an unrelated job, make it a point to show what you learned from the experience. Thinking in this way will make any job enjoyable and help you to build skills that can push you toward your goals in your engineering career.

1.7 The Importance of Honesty During the Interview Process

I recently spoke with an engineer who had just been laid off from his job and is in the process of looking for a new one. He asked me, "What should I tell prospective employers? How can I avoid telling them that I got laid off? What if I just tell them the company wasn't right for me and I wanted to find something better?"

My response was simple. I told him that you should never, ever lie during this process (both on your résumé and during the interview stage), for several reasons. Most importantly, dishonesty is the fastest way to tarnish your reputation. Being known as a liar leaves a really bad mark against your employment and personal record. Secondly, the prospective employer will most likely ask you for several references before offering the position, and hiring managers will probably want one of those references to be from your last employer. Honesty is the best policy.

Keep in mind, dishonesty is the fastest way to tarnish your reputation.

I do think that there are times when it is okay to leave something off of a résumé. For example, maybe you were laid off after working at a company for two years, and then you got a new job that lasted for only two months. In my opinion, it's fine to leave out the two-month job on your résumé as long as you are showing that the previous job ended when it did. In fact, I would highly recommend doing this, because having a 2-month-long job on your résumé looks really bad to prospective employers (unless, of course, it was an internship).

Another issue to be aware of is the validity of your LinkedIn® profile. This same individual also asked me if it was okay to leave his last position as his current position on his LinkedIn profile even though he didn't work there any longer. I told him he could probably leave this information as is for a couple of months or so after being laid off, but wouldn't recommend doing so longer than that. However, if you leave a company or were terminated, that company has the ability to request that LinkedIn edit your profile so that it shows you no longer work there. Again, you want to be smart about which actions to take or not take when creating both your résumé and LinkedIn profile so as not to tarnish your reputation.

> "Whoever is careless with the truth in small matters cannot be trusted with important matters."—Albert Einstein   [3]

1.8   Seven Steps to Creating a LinkedIn Profile That Can Land a Job

Another question engineers ask me all the time is this: "Can LinkedIn really help me get a job?" The answer is—*Yes, Yes, Yes!* However, most engineers and students don't set up their LinkedIn profiles properly allowing them to be found by potential employers.

I have more coverage about LinkedIn in Chapter 7; but as a starter, here are seven recommendations for setting up your LinkedIn profile to ensure that it helps you get noticed by employers and recruiters looking for engineers with your skills. Please note that these recommendations are more general strategies than specific technical instructions for using LinkedIn. As with all social media sites, LinkedIn changes on a daily basis, but most of the following information should be applicable regardless of the site makeup:

1. *Your title or headline on your LinkedIn profile has a huge impact on search results.* Don't just list yourself as the title on your business card (e.g., John Doe, Project Manager). Make sure your title reflects what you do at your current position. For example, your title might read: Structural Engineer, Civil Engineer, or Environmental Engineering Consultant. (Yes, you can have multiple titles in the headline.) Think about the key words people may use to search for suitable candidates when looking for someone with your skills and use them in your headline. See Figure 1.4.

2. *Fill out your profile completely.* The more words and information on your profile, the more words that can be picked up in searches. Describe all of your past jobs in detail going back to your first job. This step is very important, not just for coming

up on searches in LinkedIn, but LinkedIn profiles often come up on the first page of Google searches as well. Go ahead.

3. *Claim your public profile.* LinkedIn gives you your own public profile, but it contains randomly generated numbers. You can edit it and use your name or something close to it if your name is not available. For example, mine is listed as http://www.linkedin.com/in/anthonyjfasano. This is important, as many people believe that LinkedIn profile URLs will eventually be on everyone's business cards.

> Many people believe that LinkedIn profile URLs will eventually be on everyone's business cards. So, edit your LinkedIn URL to show your name.

4. *The section labeled "Summary" should include a few paragraphs summarizing the benefits that you can provide a prospective employer.* Don't write something that overreaches or is meaningless, such as "Highly motivated civil engineer blah, blah, blah." Instead, write, "I have a keen ability to bring in projects on schedule and under budget. In managing my last project, I was able to save our client 1 million dollars by selecting a new material to be used." Remember that employers are thinking about satisfying their needs—that's it.

5. *The section labeled "Specialties" should include all of the skills and services you are capable of providing.* I recommend listing them in bullet form or separating them with commas. For example, for a civil engineer, it might read as civil engineering, site engineering, storm water design, construction cost estimates, construction inspections, shop drawing review, client management, proposal preparation, etc. This is also very important when it comes to your qualifications and specific words being found by search engines.

6. *Seek recommendations from trusted connections, especially former supervisors.* You don't need 100 recommendations, but 3–5 solid ones can certainly provide great perspectives to a prospective employer. In seeking recommendations, ask for them in a way that will ensure they are powerful. For example, ask a potential reviewer to speak to specific qualities: "Can you please give me a recommendation? It would be really helpful if you answer the question, 'What makes Anthony a good project manager?'" Trust me; this works.

7. *As uncomfortable as it may be, adding the words "unemployed" or "in transition" to your current job title could help you land a job.* My wife did this and a recruiter found her LinkedIn profile by searching with the words "unemployed engineer." My wife ended up getting a great job with a reputable construction company, just 15 minutes away from our house.

Not only can these seven tips help you to land a job on LinkedIn, they will also help you to keep a powerful LinkedIn profile throughout your career. They are easy, and best of all, a basic LinkedIn account is free.

If you would like to see a sample of these items listed earlier, feel free to review my LinkedIn profile and connect with me: http://www.linkedin.com/in/anthonyjfasano.

FIGURE 1.4 Sample of LinkedIn profile. Here is the top portion of my LinkedIn profile as described in Section 1.8.

1.9 Your LinkedIn Profile and Your Résumé Should Be Perfect Professional Snapshots

It's funny. It seems that so many of us engineers think the same. We all look for steps or processes to guide us in everything we do. Many of the engineers I work with ask me what steps to use in preparing their résumé or LinkedIn profile: What size font? How many bullets? How many pages? What title should be used? And while there are certainly some recommended guidelines as I have already shared with you, there are no specific steps or procedures to follow when creating these items. But here is the overarching commonality: *usability.* Is it optimized for the user?

I actually believe that you should abandon processes to some degree when preparing your résumé or LinkedIn profile and be creative. Most engineers have one or more very special thing(s) about them that don't necessarily fit into a bulleted item or within one of the standard résumé sections. It is your responsibility to make sure that these unique characteristics make their way onto your résumé, some way, somehow.

As I discussed earlier, unless you have a specific job number or position, if you are going to have an objective at the top of your résumé, it should highlight what makes you different from others, not just a desire to get a job. Along the same lines, you should highlight something special, interesting, or different about you, through a sentence or two at the top of your LinkedIn profile.

For example, one of the engineers I recently coached went to school for structural engineering, but due to the sluggish job market, he was forced to take an environmental job out of school. He is now a few years removed from school and really wants to obtain a structural position, but has no experience in this area to show on his résumé. However, his father owned a structural engineering company while this individual was growing up. So I asked him to write up a few sentences about how he came from a structural engineering family and was around this discipline all his life. He added these sentences to the top of both his LinkedIn profile and résumé. Subsequently, he landed an interview and, ultimately, a position with a structural engineering firm. Another engineer I am coaching speaks four languages fluently, so I asked him to write a strong introduction paragraph highlighting this talent, as it is extremely important for international employers to be aware of this special qualification.

In addition to any unique or special traits you possess, you should also use the experience portion of your résumé and LinkedIn profile to explain the history of your career—how you got from point A to point B. Depending on the amount of room you have, consider making it flow like a story. I did so on my own LinkedIn profile, as you can see in Figure 1.5. This is a little more engaging than just listing all of the responsibilities that you had in each position. It can also help the reviewer to connect deeper with you.

So, please remember that when you are preparing your résumé and creating your LinkedIn profile, share your unique story. Just think of it this way: many engineers have performed the same tasks as you in their careers; *make sure you list how you are different from them and why prospective employers should know this key information.*

Bestselling Author for Engineers
Engineer Your Own Success
July 2011 – Present (2 years 7 months)

Once I started my company, I traveled across the US giving a talk I called "Take Your Engineering Career Wherever You Want to Take It," in which I discussed the key skills that you should develop to become a well rounded engineer.

The response to this seminar was so overwhelming that I decided to put the information into a book entitled, "Engineer Your Own Success: 7 Key Elements to Creating an Extraordinary Engineering Career." The book became an Amazon.com bestseller when it came out in July of 2011, but more importantly it gave me more speaking opportunities to get my message out to engineers all over the world.

To check out my book, please visit ► www.EngineerSuccessBook.com

Professional Speaker for Engineers
Powerful Purpose Associates
June 2009 – Present (4 years 8 months) | Ridgewood, New Jersey

Once I found success at an early age at Maser, and saw how many engineers were struggling to advance, I made it my mission to help engineers worldwide to develop their non-technical skills, creating exciting, enjoyable careers.

I founded a company called Powerful Purpose Associates (PPA) specializing in engineering career development coaching and training services. Since I founded the company, I have visited with and spoken for thousands of engineers across the world. Through PPA I founded the Institute for Engineering Career Development; a mastermind community for engineers. The IECD has since grown and taken on a life of it's own and has been transitioned into its own company.

Through PPA I still offer speaking services, book sales, and online marketing consulting services for non-engineering companies.

If you are interested in having me speak at your event for engineers, visit ► www.PowerfulPurpose.com

FIGURE 1.5 Share your story. Explain the history of your career in the experience portion of your résumé or LinkedIn profile and make it flow like a story.

1.10   Key Points to Remember

1. A winning résumé is one that helps you to land the job that you are applying for—nothing more and nothing less.

2. Recruiters and hiring managers spend about 10 seconds per résumé, identifying which ones look good enough for them to come back to for a more detailed review at a later date. The visual presentation of your résumé is critical for being chosen for a job interview.

3. It is important that you customize each résumé to match the requirements of the company. Therefore, if you are applying for 10 jobs, you should have 10 different résumés.

4. An objective is not always necessary or beneficial on your résumé. If you utilize an objective, make sure that it is unique and creative or used to identify a specific job (i.e., includes a job number).

5. If you are a recent graduate, your résumé should be no longer than one page. As you progress in your career, your résumé may be longer depending on your experience level and number of projects you have worked on.

6. Non-engineering experience can be valuable on your résumé, but you must portray it in a way that describes how this experience will benefit the prospective employer.

7. Always be honest on your résumé and in the interview. A liar carries a bad reputation that is very difficult to shake.

8. Your LinkedIn profile is essentially an online résumé that is open to the public. If you have one, be sure you spend the time to make it look professional and representative of your capabilities.

A Boost from Your Professional Partner

You shouldn't think of your résumé as a tool to show prospective employers how great you are; instead, use it as a way to show them that you can fulfill their needs. This should help expedite your job search.

Your Résumé Assistant,
—Anthony

References

[1] R. Norton, Résumés Are Dead and What to Do About It. Salt Lake City, UT: Shadow Mountain, 2012.

[2] K. Whitehorn. BrainyQuote.com, Xplore Inc, 2014. Available at http://www.brainyquote. com/quotes/quotes/k/katharinew392393.html (accessed on April 14, 2014).

[3] G. J. Holton and Y. Elkana, Albert Einstein, Historical and Cultural Perspectives. Princeton, NJ: Princeton UP, 1982, p. 388.

2

Landing and Acing an Engineering Job Interview

Opportunities don't often come along. So, when they do, you have to grab them.
                                          —Audrey Hepburn [1]

2.1 Leverage LinkedIn Groups to Land a Job Interview

In this section, I want to give you some specific strategies for leveraging this powerful social media networking tool. There are always plenty of engineers currently looking for work on LinkedIn. They spend hours searching through the site with no result to show for their efforts. The reason for this, most times, is that they are not using LinkedIn effectively or they are not very familiar with the site and how it works.

The component of LinkedIn that will provide the best chance at landing a job interview is the LinkedIn Groups functionality. There are LinkedIn Groups for every profession and every discipline of every profession. This is good, especially if you are currently looking for an engineering job. For example, if you are a structural engineer or interested in becoming one, you can join some structural engineering groups; if you are an environmental engineer, you can join a few environmental engineering groups; and so on. To take it one step further, if you are a structural engineer focusing on athletic stadium design, you can join a group for structural engineers in that arena. Yes, they get that specific.

Engineer Your Own Success: 7 Key Elements to Creating an Extraordinary Engineering Career,
First Edition. Anthony Fasano.
© 2015 The Institute of Electrical and Electronics Engineers, Inc. Published 2015 by John Wiley & Sons, Inc.

This is a great feature for many reasons. Where else can you find hundreds or thousands of employers in your specific discipline in a matter of seconds? Secondly, the Groups facilitate networking with some key people in these prospective companies including human resource (HR) professionals and hiring managers.

Here are some hints for properly leveraging the contacts within a LinkedIn Group:

- Don't start connecting with hiring managers and HR professionals as soon as you join a group.
- Don't start messaging hiring managers and HR professionals in the group asking them if they have any job openings. (Hundreds of engineers are doing that to them and you will just annoy them.)
- Don't just join the larger groups. Look for smaller ones in your niche where you can build stronger relationships.
- Don't bother people by being excessive with your posts, discussions, or messages.
- Do engage in as many discussions as possible and try to target ones where hiring managers and HR professionals are present.
- Do provide thoughtful remarks and comments in these discussions, which do not include statements such as "Is anyone hiring?"
- Do try to connect with these key people after, *and only after*, you have already been involved in a discussion with them. Use the discussion area as a common ground for establishing the connection and then start building a relationship.
- Do post a discussion saying something to the effect of "I am currently looking for a position in the XYZ field (be specific). Can anyone recommend specific networking events or professional associations where I might meet potential employers?" Ask for help, *not* a job.
- Do reply privately to people when they respond to a discussion started by you. Thank them and ask them if it would be okay to connect with them.
- Do start a group in your industry and put effort into maintaining it with good, fresh, original content. This can be a powerful networking tool for you down the road and will also be fun.

By following these simple tips, you should put yourself in a great position to build relationships with the right people during your job search. LinkedIn is a great resource *if* you leverage all of the tools that it offers. If you don't, you may just miss out on some great engineering career opportunities.

2.2 Understanding Prospective Employers and Their Needs

Between customizing your résumé (see Section 1.4) and leveraging LinkedIn Groups, your chances of landing an interview should be very good. However, once you have an interview, this idea of customization must be taken to the next step.

Once an interview is secured, your strategy shifts a bit. Now, you need to understand how you fit the need of the specific job that you are interviewing for so that you can make this information clear to the prospective employer during the interview.

Gaining an interview means the potential employer already thinks you are probably a good fit. Now, it is time to convince everyone that is true. Here's how you do it.

Review the job description in detail again. If possible, ask your contact at the prospective employer or your recruiter (if you are working with one) if there is any information about the particular position that you could review in addition to what is listed in the general job description.

Do a thorough check of your LinkedIn network and general contact database for people who:

- Work for the company you are going to interview with
- Hold similar positions to the one you are going to interview for

If you happen to know people who work for the company, contact them and ask if they know anything about the position or department. Use this information to create a plan that you could share with your prospective interviewer, showing how you would approach working in this job. There's no reason to reveal that you had coaching, unless the interviewer asks and you have permission to share your contact's name. Tell the interviewer that you have been in situations in companies with similar makeup, and describe how you conducted yourself when working for them. By doing this, you are basically showing them how you will solve potential problems in that particular position.

You might not know anyone who works for this prospective employer. In that case, through one of your contacts, try to find people who hold a similar position in a similar sized company. Ask them what challenges they face in that position and how they have successfully overcome them. Then using the information you gathered, plus incorporating your own experience and skill sets, prepare a plan that describes the steps you would take and strategies you would implement to improve the department that you would join.

> Go into the interview understanding the prospective employer's exact needs and how you can fill them.

Companies that are hiring, especially in a tough economy, have a specific need, and they will only hire someone who they believe will fill that specific requirement. In the past, a strong résumé and sharp interviewing skills may have helped to land a job with a company regardless of the specific need they had. Because if they really liked an applicant, they probably had enough business to make the hire whether it was a good fit or not. Those days are long gone. Now, you must go into the interview understanding the prospective employer's exact needs and how you can fill them.

2.3   Interview Research and Preparation

Prior to a job interview, research the prospective employer, as well as those that will be conducting the interview. You must also prepare for potential trick questions, concerns, or objections that the interviewer may pose. A prospective employer may even give you a quiz prior to the beginning of the interview. In my old engineering company, they gave a 10-question quiz on AutoCAD before the interview even started.

First and foremost, you should do whatever research is required to understand the specific need of the prospective employer. With respect to researching the company, I recommend a thorough review of its website and a Google search to find recent news about the company. If you can find out some information about a current project there, it may make for a great conversation icebreaker when the interview starts. They may ask what you know about the company, and if you can answer thoroughly to the point of actually talking about one of their projects, it may really impress them and also set the tone for a great discussion to follow. Another great tactic is to review any annual reports that the company may have filed. If a company is publically traded in the United States, those annual reports should be available for you to review. You can also do a patent search on a company to see what types of projects are in process.

Even better than just researching the company, if you know who will actually be interviewing you or who your future manager will be, you ought to try to find out as much about those individuals as possible. This is where our good old friend LinkedIn comes into play again. Do a search for these people on LinkedIn and you may be able to learn important information about their career accomplishments. You may even find that they are reading a book that you like, or have attended the same school as you, or additional points to start favorable discussions during the interview. Again, this may not only impress the interviewers but also create a powerful connection with them, which may have an impact on hiring decisions.

Once you have completed your research, spend some time preparing for the actual interview. The expected questions are not so terrible. But what can a person do when the interview starts to be more "interesting?" I have offered a few insights below to interesting questions or items posed at interviews:

1. **What are your salary requirements?**
   Avoid giving a straight answer to this question at all costs. The reason to avoid it is simple: answering lower than what the prospective employer had in mind may result in a decrease in salary that may take a few years to get back. On the other hand, giving a dollar figure that is higher than what they expected to pay may cost you the job. Part of your research should be to identify the average industry salary for the position you are interviewing; then if pushed to answer this question, give a range of +/- $5000.

2. **If you were an animal, kitchen appliance, candy bar, or others, what would you be and why?**

   This is an interesting one. The key with all of these types of questions is to give an answer that ties back into how you can help the prospective employer. For example, you might say, "I see myself as a refrigerator. I tend to be a critical component to organizations I belong to, especially in keeping things fresh with new ideas. People know they can always come to me when they need nourishment such as ideas or perspectives on how to look at things." It sounds a bit corny, but so is the question. Always bring your answers back to how you can help the prospective employer. That will resonate with them more than anything else.

3. **We are concerned that you have yet to work in this geographic region.**

   You might hear this concern if you are relocating. A good response would be to emphasize your strengths and how you feel they would still give you (and the company) an advantage regardless of the location of the company.

4. **We are a little nervous about your past employment history considering you have worked for three different employers in the past 5 years.**

   Job-hopping is always a difficult question to overcome. If this applies to you, try to explain the reason you left (or were released) from these past positions in the best possible light. Prepare an answer that will make them feel comfortable hiring you.

The most important thing to remember when facing these types of questions, objections, and concerns is to always be honest. Lying or misrepresenting yourself is only going to hurt your reputation. You could lose the position, making it even harder to get a job the next time around.

2.4   Interview Etiquette and Attire

The day has come and you are getting ready to head out to your job interview. The following are some guidelines with respect to interview etiquette and attire. I know some of these points may seem totally obvious, but you would be amazed what some people wear, say, and do during a job interview:

- *Familiarize yourself with the location of the interview well in advance of the actual interview.* Please don't solely rely on Google® maps or a $0.99 smartphone GPS app to get you to the prospective employer's office. In addition to using Google and your GPS, I recommend calling the office and asking for directions just to be safe. I also recommend doing a "dry run" drive to the office building at some point during the week prior to the interview. That may sound excessive, but if you are late to an interview, it could cost you a job. Be sure you understand where you can park, too.

- *Wear a suit. No substitutions allowed.* Please, wear a suit to the interview, no matter what advice you receive. Just because it may be a construction company, that doesn't mean you can wear jeans and work boots. When people see you for the first time, they make an instant judgment. Failure to dress properly can set you back with the interviewer(s) before you even say a word. Even if you have been told that the company has a casual culture, you are better off dressing up and making a great first impression.

- *Shut your cell phone off before you enter the prospective employer's office.* Bringing a cell phone into an interview is like carrying a loaded weapon. If it goes off, you might kill any chance that you had at landing that job. It's simply not worth it to even put it on vibrate. Just shut it off, or better yet, leave it in the car. Also NEVER play games on your phone while waiting in the office of the prospective employer. Don't laugh—I have seen it happen.

- *Show up at least 15 minutes early.* Yes, at least 15 minutes early. And when you get there, go right into the office. Tell the person at the front desk that you wanted to show up earlier in case there was any paperwork that you had to fill out. Being early can only help you when the prospective employer evaluates your overall performance.

- *Pay a compliment to the interviewer(s).* When you arrive early, ask the front desk person to point you to the restroom. On your walk to the restroom, take a look around the office. Look for project awards or announcements displayed on the walls and just observe the overall layout of the office. When you go into the interview, compliment the interviewer(s) on something that you saw such as an award that was on the wall, if possible. Do not manufacture a compliment; be sincere. If you find nothing at all that is of interest in the office, skip the compliment.

- *Do not cut people off during the interview.* You will most likely be nervous and excited during the interview, which is totally understandable, but avoid interrupting others. Make it a point to listen to their questions and respond only when they are completely finished talking. This is much easier said than done, but if implemented, this strategy will improve the connection between you and the interviewer(s) because it is a subtle sign of respect.

- *Don't ask for a drink, but don't refuse one if offered.* You don't want to inconvenience anyone, but if the interviewer offers you a coffee or soft drink, don't turn it down. If you don't drink coffee, just say, "A glass of water would be great, thank you." This again strengthens the connection with the person and gives you another minute or two to gather your thoughts prior to diving into the interview.

I am sure some of these guidelines seem obvious to you, but each one of them is critical to your success during the job interview.

2.5 Performing During the Actual Interview

A job interview, while it may only last for 30–60 minutes, could be the most important minutes of your career. When the interview starts, the interviewer(s) will most likely ask you to tell them a little bit about yourself. This is where many people flounder. Do not spend five to ten minutes telling them everything about yourself and listing every amazing thing you've done in your career. That's not what they want to hear at all. Remember: they are looking to fill a specific need. I recommend preparing a two minute introduction basically describing your background and some of your work experience that is relevant to the position.

Once you have introduced yourself, the most important part of the interview follows. At this time, you should ask them a question, and it should be something such as "What are the specific responsibilities or daily activities you would expect from this position?" Once they answer this question, you will be clear about their needs and you can explain in detail how you can satisfy them. For example, if they respond by saying that the person they hire will be responsible for doing bridge inspections and preparing construction reports, you can respond by telling them why you are a good fit for this position. You might say, "Well, that's great because in my last position I did several bridge inspections and prepared reports that were submitted to my supervisor and eventually distributed to the entire project team." Do you think that this response might impress them? Of course, it will. It may not be flashy, but it's exactly what they're looking to hear.

Your ability to express how you can fill the exact needs of an employer during an interview will dramatically increase your chances of landing that specific job. I have used this simple tip to help countless engineers ace their job interviews, and now, you can do the same.

2.6 The Follow-Up to the Interview

Inevitably, when I speak for engineers on the topics of résumés and interviews, I always get the question, "How do you recommend we follow up after the interview?"

First of all, I recommend that at the end of every job interview, you ask the interviewer(s) politely, "When you do think you will make a decision?" A typical answer is that a decision would be made in about 2 weeks.

The first thing to do after the interview is to send a thank you note via e-mail to all of the individuals from the prospective employer who were involved in the interview process. This will most likely include the HR manager and the engineering managers that interviewed you. If you are working through a recruiter, blind copy him or her on the e-mail.

Here is a brief example of what the wording might say. Change it up to make it sound like your own words.

> *I wanted to thank you for taking the time to meet with me today. I am really excited about the opportunity to work for your company as it sounds like my experience as a [fill in the blank] is a perfect match for the position. If you have any other questions or need additional information, feel free to contact me.*

You may or may not receive a response to this e-mail; however, odds are that if they are interested in you, you will hear something within a few days. If you present yourself as a match to fill their needs, they are not going to want to miss out on hiring you. If you do not hear back from the company in 2 weeks, then call your interviewer and say:

> *Hello [full name with title as appropriate]. I just wanted to follow up on our interview. You had mentioned that you expected to make a decision in about 2 weeks, so I just wanted to call and see if the position has been filled yet or not.*
> *OR*
> *Hello [full name with title as appropriate]. I just wanted to follow up on our interview and see if I could provide you with any additional files that would improve my file in the application process.*

The answers to these types of questions will most likely give you an indication of which way the company is leaning in its final decision. Sometimes, the hiring process does get drawn out because of multiple candidates or vacation schedules, so keep that in mind. If you are working with a recruiter, always coordinate all correspondence with the prospective employer through the recruiter.

So, yes, you should follow up, but be brief and polite and don't hold your breath if you do not hear back in a week or so.

2.7 Jobs Can Affect Your PE License

So you built a winning résumé, nailed your job interview, and now, you have an official offer. What's next? You take the offer right? Hold on a second—not so fast.

Something that is very important to consider before accepting a job is how it will affect your ability to obtain a professional engineering (PE) license or, if you reside outside of the United States, whatever certification your country holds for engineers (see Chapter 4 for requirements for engineering licensure).

2.8 Key Points to Remember

1. A winning résumé is one that helps you to land a job interview—nothing more, nothing less.
2. Hiring managers will typically scan your résumé for 10 seconds and decide whether or not to do a more detailed review, so be sure your résumé can capture their interest quickly.
3. Your résumé should be customized for every job application.
4. Be honest throughout the job search process.
5. You must build a strong presence and network on LinkedIn. Remember to leverage LinkedIn Groups to find new opportunities.
6. Do your research on a company prior to an interview and be prepared to discuss some aspects of that company, preferably current projects.
7. Show up on time for your interview, dress appropriately, and turn your cell phone off.

8. Be sure to ask a question about the specific duties of the job early on in the interview, so that you can present yourself as a perfect match for the position.

9. Before accepting a job offer, be sure that the experience you gain will be acceptable toward your PE license.

A Boost from Your Professional Partner

When going on a job interview, always remember that the prospective employer cares more about their needs than yours.
More than you think.

Your Interview Advisor,
—Anthony

Reference

[1] Audrey Hepburn. BrainyQuote.com, Xplore Inc, 2014. Available at http://www.brainyquote.com/quotes/quotes/a/audreyhepb413496.html (accessed on April 14, 2014).

Part II

THE 7 KEY ELEMENTS TO AN EXTRAORDINARY ENGINEERING CAREER

3

Career Goals Act as Your Destination

The future you see is the future you get.
—Robert G. Allen [1]

3.1 Career Goals Act as Your Destination

Whether you are just starting your career or have years of experience under your belt, career goals are critical to your success. When was the last time you climbed into your car and drove somewhere without knowing where you were going? When we go through our career without goals, it's like we are driving in circles not knowing what direction to take or what off-ramp to choose.

Our goals act as the ultimate destination for our career.

When pursuing a known destination in our career, we can make decisions that will get us there quickly and efficiently. Without knowing our destination, the decision-making process can be very difficult. To continue the driving analogy, if you come to an intersection, it's much easier to decide which way to turn if you know where you are ultimately going. It's one thing to take a chance and make a left turn, hoping things work out when driving on an unfamiliar road; however, when it comes to your career, *believing in blind faith is not a good strategy.*

Engineer Your Own Success: 7 Key Elements to Creating an Extraordinary Engineering Career,
First Edition. Anthony Fasano.
© 2015 The Institute of Electrical and Electronics Engineers, Inc. Published 2015 by John Wiley & Sons, Inc.

"Goals allow you to control the direction of change in your favor." — Brian Tracy   [2]

3.2   Start by Defining "Success"

Before we dive into the process of setting goals, there is one underlying goal that all professionals share and that is to be *successful*. However, they all define "success" differently. It's important that you define what success means to you as early on in your career as possible. Your definition of success will directly affect your professional performance and your life as a whole. And moving forward, you be willing to reexamine that definition of success so that it reflects the evolving "you" at the center.

It would be most helpful to clearly understand what success means to you prior to going through the goal-setting process in this chapter; however, no matter when or how you define it, just be sure to do so. Your definition of success will not only impact you as a person, but it will also affect every decision you make from that point on.

Your definition of success will not only impact you as a person, but it will also affect every decision you make from that point on.

For example, if you define success as making $150,000 a year, and you are far from that goal, then you might feel like a failure or very unfulfilled. Yet if you define success as having a job that is enjoyable and allows you to spend a certain amount of time with your family, then you might look at things from a totally different perspective. You may be making $90,000 a year, working 40–45 hours per week, allowing you to spend plenty of time with your family. If having this work/life balance aligns with your goals, then this situation might make you feel very successful.

Do not be afraid to change your definition of success as your career progresses. Many professionals start out thinking of success in terms of monetary value only, and as they grow older, they shift their definition of success toward valuing family time and decreased time at the office.

I am urging you to take the time right now to define what success means to you, and in doing so, be sure to take into account both your values and dreams. Take the time to do this now so that you don't spend years chasing what you believe to be success, only to find out that what you thought was success wasn't what you truly desired. *You can't take that time back and time is certainly a valuable commodity.* Understanding your true goals is not something you can ask other people to help you with, as only you know your innermost desires. Nobody else will be able to define what success means to you because everyone is different.

Take some time to reflect upon your past and everything that is important to you. Talk to your family members, especially your spouse, as their input may help you realize what's important and assist you in molding your unique definition of success.

```
┌─────────────────────────────────────────────────────────────────┐
│  Your Definition of Success                                       │
│  _____     │
│  _____     │
│  _____     │
│  _____     │
│  _____     │
│  _____     │
└─────────────────────────────────────────────────────────────────┘
```

Unfortunately, you will not find a class called "Defining Success 101" in your engineering curriculum or company, so it's imperative that you take it upon yourself to create this definition. This certainly can push you in the right direction, not only how you map out your career but also how you live your life. I am confident that you will be successful; however, being successful starts with first understanding what success means to you.

3.3 Define Your Values

People will tell you throughout your career that you must have goals. But it is rare to hear anyone tell you how to set them. So, let's begin that process now.

An important first step in setting goals is taking a good look at your values. These are variously defined as *principles, standards, or qualities considered worthwhile or desirable.*

What are your principles? What are your standards? What are your values? Take a notepad and paper and make a list of *everything* valued in your life. Your values might include your family, friends, honesty, integrity, freedom, or flexibility, among others.

This is a very important first step in goal setting because your values will act as the foundation for achieving your goals. You may have to dig deep to clearly identify your true values. For example, if you list freedom, ask yourself, "What kind of freedom?" Is it financial freedom, religious freedom, freedom of working for an employer? Asking these types of questions allows you to get very clear on what's important to you. Take a few minutes now to turn to Action Exercise Worksheet 11-2 in Chapter 11 and list your values. You can also download this worksheet at www.EngineeringCareerCoach.com/EYOS.

```
┌──────────────────────────────────────────────────┐
│  Sample values                                     │
│                                                    │
│      Family              Happiness                 │
│      Friends             Service                   │
│      Honesty             Respect                   │
│      Integrity           Community                 │
│      Freedom of...       Responsibility            │
│      Faith/religion      Excellence                │
└──────────────────────────────────────────────────┘
```

Don't take this step lightly. You know as an engineer how important a solid foundation is to any structure. Values are your foundation. Know them and live by them, as they will guide you in your journeys, both personal and professional.

3.4 Ask Yourself Where, Why, What, How, and Who

Once you have established your values, you can start the goal-setting process. Start by asking yourself a series of questions that start with the words *where, why, what, how,* and *who*. Try brainstorming with another person and consider answering the following questions along with any others you think of that may help you establish your goals:

Where do I want to be in my career in one year, two years, and five years?

Why do I want to be there?

What additional skills or training will I need to get there?

How can I ensure that I will achieve that goal?

Who can help me reach my goals?

How do my goals reflect my values?

The "where" question serves to help you visualize your career path and where you hope it takes you in the future. The answer to this question should be as measurable as possible. For example, is it a certain position in a company, is it obtaining your engineering license, or is it achieving a certain salary. You do not have to look years into the future; you can simply ask yourself the question, "Where do I want to be in 6 months to a year from today."

The "why" question is very important as the answer will most likely tie into your values that are your foundation. In fact, your answer to the why question will tell you what your real goal is. For example, let's say the answer to your "where" question is owning your own engineering company or becoming a partner in the firm where you currently work five years from now. Your why may be because you want to achieve financial freedom. Therefore, your actual goal is to be financially free and becoming a business owner or a partner is simply your vehicle for reaching that goal. This question cycle produces results pretty much every time it is used.

The "what" and "how" questions force you to think hard about what skills you will need to achieve your goals and what resources will help you to build them. Don't be afraid to ask others to assist you in answering these questions, especially those professionals that have successfully developed similar skills. The "how" question with regard to your values will serve to ensure that your values, which I discussed in detail earlier, remain present in your goals and your pursuit of them.

The "who" question will assist you in identifying the people that can help you achieve your goals and advance your career. We will talk more about these key people in Chapter 5 when I discuss mentoring. *DO NOT skip the who question and try to advance your career all on your own.* There are people out there who want to help you; you just have to find them and ask them. In fact, there are many people that have already accomplished goals similar to yours, so why start from scratch? Start to think about how you might answer these questions about your career. Worksheet 11-3 in Chapter 11 will help you cement these ideas by writing them out.

> *DO NOT skip the "who" question and try to advance your career all on your own.* There are people out there who want to help you; you just have to find them and ask them. In fact, there are many people that have already accomplished goals similar to yours, so why start from scratch?

3.5   More on Why

As I was writing the current version of this book, I felt the urge to add a section elaborating more on the question **why**, because it is so important to not only your success but to ensuring that you have a powerful, purposeful, and enjoyable engineering career.

If you do not question *why* you do things in your engineering career and life, then you might just do things for the sake of doing things, and not because you want to do them or because you are passionate about them. Doing things for the sake of doing them or because someone else told you to do them can cause you to be totally disengaged and disinterested in your career and life. Have you heard of those people that just go through the motions each day? They don't know their why. Have you ever asked a friend how the job was going, only to receive the reply, "It's a battle" or "Just trying to get to Friday." They don't know their *why*. It is essential to your career and your happiness to figure out your why and make sure that it agrees with your vision of success.

> If you do not question *why* you do things in your engineering career and life, then you might just be treading water. Where do you want to go, instead?

To drive this point home, I want to share with you an encounter that I had with an engineer during a coaching session that will clarify the importance of *why*. This engineer exhibited an immense amount of stress in her career and life and I wanted to help alleviate it. During one of our coaching sessions, I asked her, "What do you ultimately want to accomplish in your engineering career?" She answered, "I want to be successful and make millions of dollars so that I can donate a lot of money back to my college, so that they can continue to give students opportunities like they gave me."

I thought for a few seconds and then asked her, "Why is it important that you do this?" She replied saying that she felt so indebted to the school that she wanted to give back mainly to help students have the opportunities she did. She repeated these sentiments over and over. So I asked her, "Rather than donating millions of dollars, what other ways can you give back to the school and help the students?" She replied that she could visit the school on a regular basis and speak to the students about their career paths and answer any career-related questions that they had. In fact, she told me that she had already done this a few times. So I followed up and said, "Rather than stressing yourself out, and trying to become a millionaire philanthropist, why not just continue to speak at the school and maybe even offer to mentor a few of the students?" There was dead silence on the phone for at least 30 seconds. In that moment, she realized what she really wanted, which was to help the students. The idea of making all of this money to donate was putting so much stress on herself that she almost burnt herself out because she wasn't clear on her *why*.

Since that time, she has left her job of 10 years to take a much less stressful job, and things have been going very well for her in her career and life. Please don't put energy into chasing a goal in your career if you are not clear on why you want to achieve it.

3.6   Think Big and Then Think BIGGER!

As you go through the goal-setting process, you will find the urge to limit yourself and do what I like to call, "Thinking small."

Many professionals adopt a belief, partially due to the society we live in, that they don't deserve to be successful and achieve their dreams. Therefore, they settle for less than they deserve in their career and lives in general. Do not fall into this trap! After all, what is the point of having a dream if you can't realize it?

You are an amazingly powerful person with the ability to achieve anything you want in your career. If you don't believe this statement, you will always settle for less.

I challenge you to *Think Bigger* when it comes to goal setting and advancing your career. To start this process, insert the word "ideally" before each of the questions that you asked yourself during the goal-setting process. The following are some examples:

"Ideally, where do I want to be in my career in one year, two years, and five years? Ideally, what would I like my salary to be in two years? Ideally, what would I like my day-to-day job activities to include?" When answering these questions, imagine everything happening perfectly in your career.

What's your ideal career? What's the point of having a career that is anything less than exactly what you want to be doing, where and when you want to be doing it, with whomever you want to be doing it? You can have your dream career only if you believe you can.

> What's the point of having a career that is anything less than exactly what you want to be doing, where and when you want to be doing it, with whomever you want to be doing it?

Now, turn to Chapter 11 at the back of this book, and look for the Action Exercise Worksheet 11-3. Answer these important questions in relation to your own career development. Remember to think BIG and reach for your ideal scenarios.

3.7 Formulate and Prioritize Your Goals

If you have followed the process that I laid out thus far, worksheets 11-2 and 11-3 in Chapter 11 of the book should contain your values and goals written down. Some of these goals may be bigger than others, especially if you were thinking BIG as I asked you to.

> A study by Gail Matthews, Ph.D. at Dominican University [3], showed that people who write down their goals and commit to action items accomplished significantly more than those who did not write their goals. This study provides empirical evidence for the effectiveness of three coaching tools: accountability, commitment, and writing down goals.

Take some time to sort through all of your thoughts and formulate clearly defined goals. You can categorize and rank them however you wish. It is usually very helpful to rank your goals according to the timeline that you would like to achieve them. For example, you may set goals that you would like to accomplish six months from now, one year, two years etc. It is important to be as realistic as possible in setting your timelines to avoid setting yourself up for disappointment. To do this, check in with other professionals in your field either through social networks like LinkedIn or through professional associations, for assistance in setting realistic goals. For example, if you want to be a partner in an engineering firm, find a partner and ask how long it took, gaining insights on the average number of years in the industry to achieve this milestone.

Now, turn to Action Exercise Worksheet 11-4 in Chapter 11 of the book and list your goals and associated timelines. This will give you an overview of your goals and idea of which goals you should focus on more now based on your timelines.

Congratulations. You have just set some goals for your career. These goals will help immensely with your day-to-day decision making, which will guide you toward creating a successful career. Years from now when you look back on your career, you will be grateful that you took the time to go through this process.

3.8 Be SMART and Use Small Steps for Big Results

Now that you have clearly defined career goals, you can set out to start achieving them. One big roadblock professionals hit when trying to achieve their goals is getting overwhelmed because their goals seem so big (which they should be if you followed my advice). Many professionals don't know where to begin, and therefore, they don't start at all. Sound familiar?

One way to avoid this overwhelming feeling is to break down your goals into smaller, achievable steps. To do this, you can utilize an acronym used by many people called **SMART**.

It is best to break down your goals into smaller, achievable steps using the SMART acronym:

S—Specific
M—Measureable
A—Achievable
R—Relevant
T—Time bound

Many people like to use this process to actually set goals, but I prefer to use the method we just walked through because I believe it helps you look at the big picture and tie your values to your goals. I recommend using the **SMART** process for breaking down your bigger goals into smaller stepping stones. Here's what this acronym stands for:

S = Specific
Break your bigger goals down into smaller specific steps. For example, let's say you have a goal of becoming a licensed professional engineer (PE) in two years. Break that goal down into specific steps. For example:

- Step one may be to obtain the application form and familiarize yourself with the process.
- Step two may be to prepare a detailed schedule of what has to be submitted and when.

Doing this will not only put you in a great position to achieve your goal on time; it will also relieve those feelings of being overwhelmed because you will know exactly what to do next.

M = Measurable
Create criteria that you can use to easily measure your progress toward a goal. For example, perhaps you have a goal of obtaining a master's degree over the next two years and that it will require you to take 10 classes. One of your criteria may be that you want to finish five classes each academic year.

Establishing criteria will help you track your progress very easily. Continuing with the example earlier in this section, if at the end of the first year you have only taken three classes, then you can plan accordingly and increase your workload to meet your deadline. Or perhaps you decide to adjust the deadline by moving it back 6 months to make the goal more attainable. Either way, establishing criteria will allow you to stay on track toward accomplishing your goals.

A = Achievable

Remember in Section 3.6 when I asked you to think big when setting your goals? Please, take that advice to heart. However, when creating these steps to help reach your goals, you want to ensure that the steps are achievable. For example, you may have a goal of being the president of the company you work for in 5–10 years from now. That is an admirable and attainable goal. To get there, you will need to break the goal up into specific and measurable steps that are achievable and realistic. For example, one step along the way may be to obtain a master's degree.

This process may sound simple, but if you set achievable goals, you will accomplish them. If you create steps that are too difficult to handle all at once, you may give up in defeat and fail to achieve your ultimate goal. *You always want to put yourself in a position to succeed.*

R = Relevant

When creating these minigoals, or stepping stones, be sure that the steps are relevant to your ultimate goal. Using the example earlier of becoming the president of the company where you are employed, you may come to understand that one of the steps that you have listed is pursuing a Ph.D. Realistically, do you need a Ph.D. to become president of the company? It probably depends on the company policy, but if you are working for a consulting firm, you may not need one.

We often add steps into the process that are not relevant and actually take us off the most direct path to achieving our goal. To avoid this happening to you, choose only the steps that you absolutely need to take to achieve your goal. If you have a good mentor, which I'll discuss in Chapter 5, that person will help you choose the right steps needed to reach those goals without wasting time and other resources.

> We often add steps into the process that are not relevant and actually take us off the most direct path to achieving our goal. To avoid this happening to you, choose only the steps that you absolutely need to take to achieve your goal.

T = Time Bound

Your goals and minigoals should have deadlines associated with them. We all need deadlines in our career because—let's face it—we are so busy that if we don't hold ourselves accountable, we may never accomplish our goals. Time deadlines are a great way to instill accountability. This is very important because not only does it give you a deadline on paper, but your mind will subconsciously process that completion date and start creating an urgency to help you move toward it.

When preparing a plan to achieve your goals, be **SMART** and create specific, **m**easurable, **a**chievable, **r**elevant, and **t**ime-bound steps. *Remember: small steps bring about BIG results.*

Now, turn to Action Exercise Worksheet 11-5 in Chapter 11, where you can walk through the SMART process for your goals, one at a time.

3.9   Let Your Definition of Success Guide You

If you are following the path presented in this chapter, you have already defined what success means to you; made a list of your values; set goals based on your visions of success, your values, and your why; and then used the SMART process to prepare a plan for pursuing each goal. However, as an engineer, you know that things don't always go as planned, and therefore, we must be prepared should problems arise. Inevitably, something will happen in your career that will either slow you down or prevent you from achieving one or more of your goals.

If (or I should say when) this happens to you, the best thing to do is to fall back on your definition of success. Your goals may change over time, but your definition of success will most likely remain the same if you define it in a way that considers your true desires. If you have done a good job of defining "success," then when you falter or become stagnant in the pursuit of your goals, you can review your definition and use it to inspire you; it will also help you to determine what you should be focusing on in your career and life at that time in order to achieve your vision of success.

This point can be difficult to grasp, so let me share a personal example to make it clear. When I define "success," I take three steps:

1. First, I define what success means to me in my career.
2. Second, I write out what it means to me personally with respect to my family life.
3. Last, I combine the two into an overall vision of success.

My definition of success as it pertains to my career is as follows: to inspire as many engineers as possible to create extraordinary engineering careers and lives. My definition of success relating to my personal desires is to "be able to spend as much time with my family as possible, including traveling with them on a regular basis."

When I merge my two statements, my overall definition of success would be something like this: "I want to inspire as many engineers as I can on a regular basis while being able to spend as much time as possible with my family, doing things we like to do."

> *My Definition of Success*
> I want to inspire as many engineers as I can on a regular basis while being able to spend as much time as possible with my family, doing things we like to do.

Crafting a statement like this can help you ensure that you live a life that is extremely fulfilling and also keep you on track when you lose your focus. I even dug a little deeper and created a definition of success that is a little broader and gives me a little more flexibility with some of my career decisions: my definition of success is to always have the ability and freedom to do whatever I want, whenever I want to do it, with whomever I want to do it. While this might seem broad, it is a big picture goal, and there are many subgoals that help to make it a reality. If I lose interest in a goal, I refer to my definition and make sure that my goals are still aligned with my vision, and if not, I make the necessary changes to reengage and reinvigorate my passion and engagement. While this may seem like a broad definition, remember that it is a definition or vision of success and it is meant to be used as a guide to keep you on track. It doesn't have to be very specific.

For example, in my case, if someone were to offer me a job where I have to be in the same location every day all day, it's easy to see that this opportunity does not fit my definition of success. However, running my own business, where I have the flexibility to take on whatever clients I would like and also have the opportunity to work whatever hours I need to, to work around my kids' activities, fits well into my definition or vision of success.

Now, revisit your definition of success that you wrote earlier in this chapter and ensure that it is defined in a way that will act as a guide for you throughout your career and life. Take a few minutes to list in the box provided here your personal, professional, and overall visions of success statements.

> Vision of Success Statements
> *Personal:*
> *Professional:*
> *Overall vision:*

3.10 Motivate Yourself to Pursue Your Goals

After revisiting your definition of success and confirming that the goals are the right ones, you might be lacking motivation around pursuing some of these goals. Here's a way to jumpstart the process: try *listing the benefits associated with each goal.* When you do this, take a really deep look at the benefits. Maybe one of your goals is to obtain your PE license, but you can't seem to get motivated enough to study. Instead of pursuing the goal haphazardly, list the benefits that obtaining the license would provide such as a possible increase in salary.

But don't stop there. What would that increase in salary do for you? Maybe it would help you to purchase a new car, a house, or pay off existing debts. This process will show you how much of an impact achieving your goals might have on you both personally and professionally. This revelation should bring you the motivation that you were lacking.

It's amazing how powerful a process like this can be. When we realize the impact that our professional goals can have on our personal lives, we tend to flip an internal switch that unleashes the giant inside of us, which relentlessly pursues that goal.

"Deep within man dwell those slumbering powers that would astonish him, that he never dream possessing; forces that would revolutionize his life if aroused and put into action." —Orison Swett Marden [4]

3.11   Time to Celebrate!

You have defined success, set goals, put a **SMART** plan into action, and motivated yourself. When you succeed and achieve one of your goals, what's the next step? Celebrate! This is an important step that many people forget to do.

It most likely took quite a bit of time and effort to achieve your goal, so take a moment to celebrate your achievement. Treat yourself to a nice dinner with your spouse, a friend, or other family members and reflect upon what you have achieved.

Be sure to call all of those that might have helped you to achieve your goal and express your gratitude. Then after taking some time to celebrate, you can go back to step one and start the goal-setting process again to get to the next level on your success ladder.

Congratulations on your achievement, but remember that *while it is important to have a destination, it's also important to enjoy the journey, and setting goals that are truly meaningful to you is the best way to do so.*

3.12   Key Points to Remember

1. Career goals provide you with a destination for your career. When pursuing a known destination, you can make decisions that will get you there quickly and efficiently.
2. Before setting goals, define what success means to you. Consider creating a definition both for your professional and personal endeavors and possibly combining them into one overall vision.
3. Start the goal-setting process by taking a good look at your values, as they will act as the foundation for your goals. Your values might include your family, friends, honesty, and integrity (just to list a few).
4. Once you have established your values, ask yourself a series of questions that start with the words where, why, what, how, and who. Insert the word "ideally" before each question to ensure that you are thinking BIG. Answer these questions in a brainstorming format, writing down everything that comes to mind.
5. Sort through all of your thoughts and formulate clearly defined goals. Use the **SMART** process to break down your goals into smaller stepping stones. You now have career goals and achievable steps to reach them. Lean on your definition of success should you get off track in the pursuit of your goals.
6. When you arrive at difficult decisions along your career path, let your definition of success guide you toward the option that works best for you.
7. If you are having trouble getting motivated, list the benefits associated with reaching the goal. Be sure to look at the benefits both personally and professionally.
8. When you achieve a goal, take time to celebrate your accomplishment before you start setting new goals. An important part of career advancement is enjoying your journey.

A Boost from Your Professional Partner

When it comes to your career goals, aspirations, and endeavors, don't let others paint the picture for you. It's your canvas, your paint, and your brush.
Your canvas is HUGE—go wild; the more colors, the better!
Take ownership of it!

Your Professional Partner,
—Anthony

References

[1] R. G. Allen, Motivation Point. Available at http://getmotivation.com/robert-allen-hof.html (accessed on April 14, 2014).

[2] B. Tracy, BrainyQuote.com, Xplore Inc, 2014. Available at http://www.brainyquote.com/quotes/quotes/b/briantracy383370.html (accessed on January 30, 2014).

[3] G. Matthews, Summary of Recent Goals Research [Online]. Available at http://www.dominican.edu/academics/ahss/undergraduate-programs-1/psych/faculty/fulltime/gailmatthews/researchsummary2.pdf (accessed on July 13, 2014).

[4] Y. Hirsch, The Capitalist Spirit: How Each and Every One of Us Can Make A Giant Difference in Our Fast-Changing World. Hoboken, NJ: John Wiley and Sons, 2010, p. 287.

4

Obtain Credentials That Will Help You to Reach Your Goals

You don't win the U.S. Open by being lucky, throw it up there and hope. He's got credentials.

—Hale Irwin [1]

4.1 Credentials Bring You Credibility

Your credentials represent measurable accomplishments in your career, which include degrees, licenses, certifications, and any awards you may have won. Credentials are essential for several reasons, but the most important thing that credentials bring you is credibility. While you can certainly have credibility without credentials, in the engineering world, credentials like the professional engineering (PE) license give you instant credibility.

Whether you are trying to secure a job with a new employer or obtain a new client for your company, the more credibility you have, the easier it will be. This seems like an obvious point, but too many professionals fail to recognize the importance of credentials, and they don't pursue them. This, in turn, drastically impedes their career development.

Credentials will not necessarily make you a great design engineer or project manager. However, they will put you in a position to effectively display your engineering and

Engineer Your Own Success: 7 Key Elements to Creating an Extraordinary Engineering Career, First Edition. Anthony Fasano.

management talents. It is not uncommon for engineers to get promoted over other engineers simply because they have their PE license, when in reality the engineer that was passed over is more talented. While this behavior by companies may seem unfair, I mention it to emphasize that ultimately it's your responsibility to obtain the credentials that will help you advance in your career. Failure to do so can set you back in your engineering career, fair or not.

> It's your responsibility to obtain the credentials that will help you advance in your career. Failure to do so can set you back in your engineering career.

Not excelling at examinations and having no time to pursue a degree are poor excuses. *You are a smart, powerful individual who can do anything when you put your mind to it.* You can, and will, pass any exam that you properly prepare for in advance. This chapter will provide some recommendations and strategies on how to go about obtaining credentials that can put you in a position to succeed.

4.2   Set Yourself Apart From Others

An important part of advancing your career is setting yourself apart from other professionals. Standing out as an above-average engineer will greatly facilitate your advancement.

Credentials will certainly help you do that. When I say "set you apart," I don't necessarily mean it in a negative or competitive way. I truly believe that if you follow the guidelines provided in this book and are passionate about what you do, you will never have to worry about competition in your career. I don't want you to adopt the mindset that you always have to beat everyone else, because that's not what career advancement is all about. *Setting yourself apart means standing out from the crowd in your own way.*

> I truly believe that if you follow the guidelines provided in this book and are passionate about what you do, you will never have to worry about competition in your career.

Setting yourself apart becomes even more important when the demand for engineers goes down during bad economic times. Your credentials can help you to keep your job in tough times and secure a new position rather quickly if needed.

Imagine you are unemployed and you submit your résumé to a company for a position as a design engineer. That company receives 10 résumés for the position and decides to interview only 4 of the 10 candidates to narrow down the field. Assuming the company's hiring managers don't know the candidates personally, how are they going to decide which candidates to interview? They are going to look at the credentials on the résumés. If only four of the candidates have their PE license and you are one of them, do you think you will get an interview? I would wager that you will be on the call list. Even though it is not guaranteed, your PE license will put you in a much better position to be included in the initial interview process. Other factors that may be considered are your activities in professional organizations and your current client relationships, both of which we will cover in Chapter 7 on networking and building relationships.

Regarding the PE license, some disciplines of engineering don't require or recommend that the PE license be obtained for those individuals working in that field. For example, in the field of civil engineering, your PE license is an absolute requirement for significant advancement, whereas in the field of chemical engineering, it is very rare that these engineers obtain the PE license. Due to the large number of different disciplines, I will not discuss each one in this book, but I strongly recommend that before you pursue licensure in your field, contact an experienced engineer in your discipline and ask that person whether or not the PE license is necessary for your field. Even if only a small number of engineers in your discipline have a PE license, then pursue it. You never know where your career is going to take you and when the PE license may become valuable to you.

| "I got my start by giving myself a start."—Madam Walker   [2]

Another scenario where your credentials may come into play is when your supervisor is considering raise or promotion. It is one thing if you have had a good year and performed great work for the firm; however, it's another thing if you also obtained an advanced degree or even your PE license. These credentials give your company a concrete way to measure your development as a professional engineer.

In your career as an engineer, it's up to you whether you want to be lost in the crowd or rise above the rest. Credentials will help you to earn credibility and set you apart from everyone else.

4.3 Recognizing the Difference Between Patience and Procrastination

The biggest obstacle that engineers face in obtaining credentials is often their own procrastination. Many engineers try to cover up their procrastination by saying, "I am waiting for the right time to take the exam." What exactly does that mean? Here's a little secret: *it will never be the right time to take a course or exam in your career.* The timing depends on how badly you want this credential.

| Here's a little secret: it will never be the right time to take a course or exam in your career. The timing depends on how badly you want this credential.

In fact, it's better to tackle these challenges as early on in your career as possible. As you get deeper into your work life, you will have more responsibilities, such as managing staff, budgets, etc. Along the same lines, as time goes by, you will most likely assume more personal responsibilities such as raising a family, owning a home, and getting involved in your community. All of these factors are sure to make it more challenging to obtain the credentials that you desire later in your life.

So, the next time you find yourself waiting for the right time to pursue a new credential, make right now the right time and go after it.

4.4   Exam Preparation: Start With the End in Mind

I have taken more exams than I care to admit in my career, and in this section, I want to share a preparation strategy with you that has worked for me every time. When you are pursuing a credential that requires you to take an exam, the first thing I recommend doing (after you get approved to sit for the exam) is to prepare a study schedule. Figure 4.1 outlines the typical times that I would use to study when preparing for the PE exam. In my case, early morning was my most productive study time; however, you will have to determine your best time to study based on your situation.

In preparing the schedule, the first step should be to place the date of the exam on your calendar and then create the study schedule working backward. Based on the number of weeks until the exam and the number of topics you have to prepare for, you can come up with a fairly detailed schedule. Be sure to include the schedule as part of your day-to-day calendar; do not just write it down on a loose piece of paper to lose among the clutter on your desk.

> Your study schedule should be on your calendar with all of the other important events in your life.

Professionals too often seem to treat exam preparation as secondary and push it aside when anything else comes up. If you find yourself doing this, stop and think for a few minutes about the importance this credential has on your career. Think about how it will impact your career development as well as any benefits it might have on your personal life (see Section 3.10 for other ways to motivate yourself and regain your focus).

Studying for an exam whenever you have the time to fit it into your schedule is not a good strategy, and professionals that I have spoken to that employ this method fail the

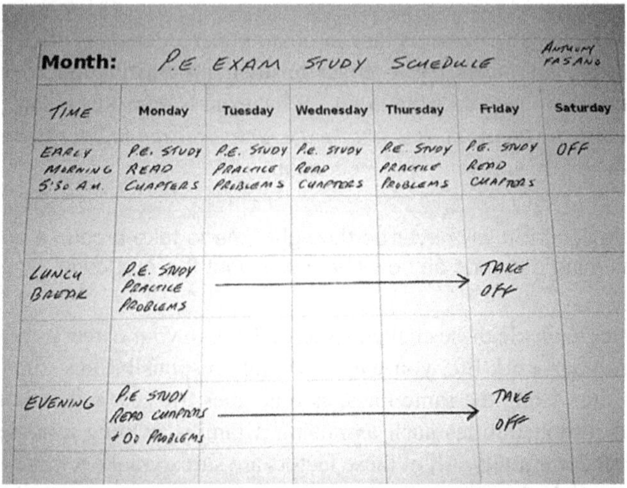

FIGURE 4.1 PE exam preparation schedule. A strong, detailed plan for tackling exam preparation is essential for success.

exam more times than not. If you can stay late at work right now to finish a project, why can't you stay an extra hour to study for the exam? I have had success studying early in the morning for an hour or so before I left for work. It means you will be getting up earlier, but the commitment is necessary. No matter how early I woke up, that extra time was always scheduled on my calendar as study time. *Remember, a few months of hard work can bring you a credential that will impact your career forever.*

> Studying for an exam whenever you have the time to fit it into your schedule is NOT a good strategy, and professionals that I have spoken to that employ this method fail the exam more times than not.

4.5 Tips for Approaching the PE Exam

While the information provided in Sections 3.10 and 4.4 of this book will help you to prepare and motivate yourself for an exam or any other challenges you will take on, I wanted to dedicate a section of the book specifically to the PE exam, because this test seems to be one of the biggest obstacles in the careers of engineers. Its official name is "The Principles and Practice of Engineering exam"; however, most engineers refer to it as the PE exam.

The Principles and Practice of Engineering exam is given only in the United States and its territories, but many countries have a similar exam. Please understand, I have only had experience with the US version of the exam. However, the fundamentals of this chapter should help you prepare for the exam regardless of where you reside. In Section 4.5.10, I have summarized the credentialing process for 11 other countries with large engineering populations.

> Whether you follow the steps in this section or not, remember that the PE license is the most important credential that you will pursue in your career as an engineer. It has the potential to make or break your career.

4.5.1 Take the Fundamentals of Engineering Exam as Soon as Possible

As of the time this book was published, engineers in the United States are required to pass a six hour computer-based and an eight hour pencil-and-paper exams in order to obtain the PE license. The first exam is called the Fundamentals of Engineering (FE) exam. Again, this specific exam is administered in the United States. For licensing requirements in other countries, visit the websites provided in Section 4.5.10. The FE exam includes questions on a wide array of topics including, but not limited to, mathematics, engineering probability and statistics, chemistry, computers, ethics and business practices, engineering economics, engineering mechanics (statics and dynamics), strength of materials, material properties, fluid mechanics, electricity and magnetism, and thermodynamics [3].

The requirements for sitting for this exam are always subject to change, so be sure to check with the National Council of Examiners for Engineering and Surveying (NCEES) for the most up-to-date information. The NCEES is a nonprofit organization that administers both the FE and PE exams. More information on this organization as well as the exams can be found on the website at www.ncees.org [4]. Currently, the exam is open to anyone with a degree in engineering or a related field or anyone who is currently enrolled in the last year of an Accreditation Board for Engineering and Technology (ABET)-accredited engineering degree program. Be sure to check with the NCEES if you have questions specific to your situation.

The bottom line is this: *take the FE exam as soon as you are eligible.* Many colleges have the exam as part of their curriculum and engineering students take the exam in their senior year. When I was in college, they provided transportation for us to the exam site.

It's important to take this exam as soon as possible because it encompasses many different topics. The further removed from college you get, the harder it is to remember all of them. Many of the engineers who I know that don't have a PE license attribute it to waiting too long to take the FE exam. Either they failed the exam in their senior year of college and waited too long to retake it, or they just put it off until later in their career and then couldn't pass it. Please, take this exam as soon as possible.

> Many of the engineers that I know who don't have a PE license attribute it to the fact that they waited too long to take the exam.

4.5.2 Start the PE Exam Application Process as Early as Possible

In order to sit for the PE exam, you are required to have a certain number of years of experience (this varies by state). Engineers always seem to want to wait until they are close to fulfilling the experience requirements before they fill out the application to sit for this important exam. There is no need to wait. I recommend that preparing the application be an ongoing process that you commence as soon as you start your engineering career. You can update it monthly or every time you work on a new project.

There are several reasons for taking this approach. First of all, if you wait three or four years to fill out the application, you may have trouble remembering what projects you worked on, especially if you change employers. Secondly, by starting the process now, you will save a huge amount of time from not having to go back and retrace everything that you had done previously. Last, you will also find that by keeping the information up to date in an organized manner, you will be able to use it in preparing other documents such as professional résumés and proposals.

It is also important to keep in mind that you will need to get signatures from all of the professional engineers that supervised you on the projects that you referenced on the application. Starting the process early will give you plenty of time to track down those signatures, especially if you no longer work with that supervisor. Even if it is not time to obtain signatures, it doesn't hurt to send them an e-mail asking them if they would sign your application when the time comes so that they won't be surprised at the request later.

I have actually heard of engineers whose supervisors weren't comfortable signing for them. If this is the case, you have three options:

- First, sit down with your previous supervisor and review your previous responsibilities and accomplishments that were made while working on his or her project(s). Supervisors may forget what you accomplished because it was so long ago, but after a face-to-face conversation, they usually feel more comfortable vouching for your experience.
- Second, try to find another supervisor who has a PE license and who worked on the same project as you and ask that supervisor to sign your application.
- If all else fails, substitute another project in its place on your application.

The PE exam application may be the most difficult part of the licensing process. Begin it as early as possible to ensure a timely submission and avoid procrastination.

4.5.3 Submit the Application as Soon as Possible

After you have prepared your application using the tips in the last section, submit the PE exam application as soon as you have fulfilled all of the requirements. Many engineers wait an extra six months or so just in case the board doesn't accept all of their experience, but I believe this is the wrong way to go about it.

When you delay your submission, you leave yourself open to more procrastination, which could delay the submission even longer if other circumstances come up in your life. Submit your application and let the board make its decision: don't assume you know how the board is going to respond. I submitted my PE application at the age of 23, and I included all of my summer experience in college. Many people told me not to submit that early because the board would never accept it. I decided not to listen to them, submitted the application, and was elated when it was accepted.

Even if the board rejects your application and requires another 6 months of experience, all you will most likely have to do at that time is submit any additional experience you may have gained, as the board will most likely honor the rest of the application. There is absolutely no reason not to submit early as long as you meet the requirements.

4.5.4 Don't Take the Exam Just to See What It Contains

I have had many engineers tell me that they aren't studying for the PE exam the first time around because they are taking it just to "see the questions." This is my favorite excuse that I have heard for not studying or doing any advance planning.

Are you serious?

Maybe you want to sit and take an exam for eight hours just to look the contents over, but I'll pass on that waste of time, thanks.

Believe me, you only want to take this exam once. You are going to need to put in a solid three to four months of studying, whether you have seen the exam or not. So, do yourself a favor: make it a goal to study hard and pass the exam on the first shot. You know you can do this, so start the preparation process now.

4.5.5   Take a Review Course Whether You Want to or Not

Many engineers don't consider taking a review course for the PE exam because they are confident in their own studying habits. Regardless of how well you study, I highly recommend that you take a review course. They are fairly inexpensive and many employers will reimburse you for the fees.

If nothing else, a review course will force you to maintain a regimented study schedule. These courses usually cover a few topics during each session and may have assignments to complete. As much as we like to think we can handle this exam on our own, unfortunately, with everything that is going on in our lives, the exam preparation usually takes a back seat. The structure of a review course will help you to maintain your study schedule.

> The structure of a review course will help you to maintain your study schedule.

In addition to the schedule, these courses usually offer some really practical tips, recommendations, and strategies for dealing with certain types of problems on the exam. Many courses will even provide handouts that you can utilize on certain types of exam questions that will not only help you to solve the problem, but help you to do so faster and save time.

Another benefit to these courses is the relationships you will build with your classmates. They will become a support group and you may find yourself contacting them outside of the course to get together to study or just ask questions. This was one part of the review course that I really enjoyed. It made me realize I wasn't the only person trying to maintain a full-time job, juggle my personal responsibilities, and pass a challenging exam at the same time.

So, if you think a review course is not for you, please rethink this decision and consider taking one. You will be very happy you did.

4.5.6   Ask Others What Worked for Them

You will probably see this advice one hundred times in this book, but when you are advancing your career, there is no need to reinvent the wheel. Ask people who have faced similar challenges how they overcame them and save yourself valuable time.

This goes for the PE exam as well. Many of the strategies that I utilized in preparing for and taking the exam came from colleagues who had already passed the exam. I recommend having a discussion with those you know that have passed the exam and ask them for pointers on how they did it. While you may be able to pick up some helpful tips from those who have failed the exam, there is a reason they failed and you certainly don't want to adopt any habits that might prevent you from reaching your goal.

Most engineers who have already passed the exam would love to help you do the same. They may end up giving you a strategy that can very well make the difference between passing and failing.

> Most engineers who have already passed the exam would love to help you do the same. They may end up giving you a strategy that can very well make the difference between passing and failing.

4.5.7 Bring the Right Materials to the Exam

You are allowed to bring as many books to the PE exam as you wish, which is why on the day of the exam people will literally show up with wagons full of books. While I understand it can be tempting to have every possible piece of information with you, you must be careful not to place too much emphasis on your reference materials. The most important aspect of passing this exam is budgeting your time. You will have to answer 40 questions in each of the two four-hour sessions, so you will need to be judicious with your time.

My point is that the more books you plan to use, the more time it will take you to browse through them searching for the answer. This takes away from the examination time resulting in answering fewer questions. My recommendation is to purchase an engineering reference manual for your discipline and do all of your studying using this guide. Michael Lindeburg is the author of many good engineering reference manuals [5–8]. Taking this approach allows you to really get to know one complete manual that you most likely will be able to use to answer 75% or more of the exam questions.

You can still bring all of your other books and have them there just in case you need them. However, the reference manual approach should save you a good amount of time and also take away that feeling of constantly being rushed and not knowing where to find pertinent information.

Be sure to flag all the pertinent sections of your reference manuals with tabs. For example, if there are certain charts or graphs that you will have to use throughout the exam, place a tab on those pages and label the tab "XYZ chart." This will save you a tremendous amount of time. If you do not flag the appropriate sections of your books, you may find yourself wasting time flipping through pages instead of solving problems.

Of course, you will need a calculator for the exam; however, the NCEES only permits you to use certain types, so be sure to check the website for the list of approved ones.

Also, make sure that you bring a dictionary to the exam. That's right—a good old-fashioned dictionary. Someone told me to do this and I answered two questions correctly because of this excellent advice. Many times an exam question will simply ask for the definition of a word. All you have to do is look it up in the dictionary.

So, when collecting your books for the exam, bring what you feel is necessary to succeed, but keep your pile of reference books to a minimum.

Reference Guides and Manuals for PE Exam

FE Review Manual: Rapid Preparation for the Fundamentals of Engineering Exam by Michael R. Lindeburg, PE [5]

Civil Engineering Reference Manual for the PE Exam by Michael R. Lindeburg, PE [6]

Mechanical Engineering Reference Manual for the PE Exam by Michael R. Lindeburg, PE [7]

PE Chemical Sample Questions and Solutions by NCEES [9]

Electrical Engineering Reference Manual for the PE Exam, 5th Edition by Raymond B. Yarbrough, PE [10]

Mike's Civil PE Exam Guide: Morning Session by Mike Hansen, PE [11]

Principles and Practice of Engineering Architectural Engineering Sample Questions and Solutions, Second Edition by Mark McAfee, PE [12]

PE Sample Questions and Solutions: Electrical and Computer Engineering (Book & CD-ROM) by NCEES [13]

PE Civil Sample Questions and Solutions by NCEES [14]

Civil Engineering All-In-One PE Exam Guide: Breadth and Depth, 2nd Edition by Indranil Goswami, PE, PhD [15]

Build With Steel: A Companion to the AISC Manual by Paul W. Richards, PE [16]

2012 International Building Code by International Code Council [17]

Engineering Unit Conversions by Michael R. Lindeburg, PE [8]

Water Resources and Environmental Depth Reference Manual for the Civil PE Exam by Jonathan Brant, PhD, Gerald J. Kauffman, MPA, PE [18]

Environmental Engineering Solved Problems by R. Wane Schneiter, PhD, PE [19]

4.5.8   The Day of the Exam

The day you have been preparing for has finally arrived. You will most likely be very tired in the morning because your nerves may have kept you awake the previous night. No worries. Get up early, have a really good breakfast, and head out to the exam. If you can, it is best to stay overnight near the place where the exam will be held so you don't have to deal with packing all of your books, traveling time, and parking the next morning.

While you shouldn't necessarily study on the morning of the exam, review your material to make sure you know where all of the important tables, graphs, and other pertinent information so that you can access them quickly.

Remember that your biggest obstacle in passing any exam is time.

As of this writing, each of the two four-hour sessions is comprised of 40 multiple-choice questions. I recommend that when each session begins, first, go through and answer all of the questions that are easy for you (the ones you are 100% certain you can complete, which will probably be about half of the 40 questions). Then go back and answer the ones that you would measure as being of medium difficulty—those whose answers you're not 100% sure of, but you think you can figure out or get close to completing. If you look at a problem and have absolutely no idea of how to get the answer, skip over it and save it for last. Once you have finished the easy and medium ones (which will be about 30–35 of the 40), go back and do the best you can with the hard ones.

The message here is simple—be sure to get the ones right that you are comfortable answering. The last thing you want to do is spend precious time trying to figure out a hard problem and then not have sufficient time to answer three or four easy questions. *It's about getting as many questions right as you can in the allotted amount of time. Never lose sight of this goal.*

4.5.9 The Day After the Exam

The exam is usually given on a Friday, so make it a point to relax the next day. Your mind and your body have been pushed to the limit preparing for the exam. Give both the rest they deserve.

4.5.10 Credentialing Processes Around the World

This section provides resources where you can obtain information on credentialing for countries with large populations of engineers. Please note that this information is always subject to change, so please confirm the proper process with the proper credentialing organization in your country.

Australia	http://www.engineersaustralia.org.au/nerb/professional-engineers
Canada	http://www.engineerscanada.ca/frequently-asked-questions
China	http://english.cast.org.cn/n1181872/n1182065/n1182088/46506.html
Germany	http://www.vde.com/de/Karriere/Beruf-und-Arbeitsmarkt/Seiten/RegulationoftheEngineeringProfessioninGermany.aspx
Hong Kong	http://www.erb.org.hk/notestoapplicants.htm
India	http://www.ieindia.org/PDF_IMAGES/PDFFORM/PEGuidelines.pdf
Japan	http://www.engineer.or.jp/c_topics/000/000345.html
Korea	http://www.kpea.or.kr/english/pe_examination.html
Singapore	http://app.peb.gov.sg/pe_reg_intro.aspx
Switzerland	http://www.reg.ch/en/informationen/eintragung/
Taiwan	http://wwwc.moex.gov.tw/english/content/wfrmcontent.aspx?menu_id=512

4.6 If You Fall Off the Horse, Get Right Back On

When pursuing a license or certification, if you do not pass the exam, it is very important that you immediately sign up to take the exam again the next time it is administered. This is one of the biggest obstacles for professionals because they equate failure to pass the exam with the end of the process. It's not the end of the world—it's just one experience that you will learn from. You will utilize that knowledge in your next attempt.

> When pursuing a license or certification, if you do not pass the exam, it is very important that you immediately sign up to take the exam again the next time it is administered.

Many engineers do not pass the exam the first time around and then they delay retaking it. This postponement can go from months to years, and by the time they retake the test, the material on the exam is no longer fresh in their minds and the exam becomes even more of a challenge.

This habit of procrastination will have a negative effect on you, not only in test taking but in your career as a whole. *When something doesn't go exactly as planned, the way you react will define your career.* When faced with adversity, you can give up on your goal, or you can dig down a little deeper and get right back to chasing it. The choice is yours. I urge you to choose wisely because your decision may very well be the deciding factor in whether or not you craft a successful and enjoyable career or suffer from the "I wish I had…" syndrome later in life.

4.7 Master's in Engineering or Business Administration?

One of the questions that I get asked most by engineers is whether they should get a Master of Engineering (MEng) degree or a Master of Business Administration (MBA) degree. I will offer some of my thoughts on this topic in this section; however, I need to make two very clear caveats up front. First, this decision is extremely specific to your career path and interests and is not likely one that you can simply get answered by reading a book. Second, there are new programs every day that are being created, and existing programs are constantly being modified, so the information in this section could easily be out-of-date by the time you read this. Please keep these points in mind.

One degree that many engineers pursue is a Master of Science in Engineering (MSE). Many MSE programs do strive to prepare engineers for project manager responsibilities; however, of the degrees I discuss in this section, the MSE probably will have the most technical content. This degree focuses primarily on technically related courses and may require the completion of a thesis to earn the degree. It is important that you check with the governing organizations in your area with respect to educational requirements for the PE license. At the time this book was published, the NCEES Model Law requires a master's degree or equivalent to sit for the PE exam starting in 2020. The master's degree must be from an EAC/M-ABET-accredited program and with a specific record of three years or more of progressive experience. The legislature and/or Registration Board in each state will decide whether or not to adopt these provisions [20].

In every other scenario, unless of course the MSE is required for licensure, I would recommend a degree that incorporates aspects other than solely technical courses and information, and here is why: if you are planning to enter into the engineering industry, developing your core skills, management knowledge, and leadership abilities will help you tremendously, and doing so early on in your career through an educational program will give you a distinct edge. Not only will these skills help you succeed on a daily basis, but having them on your résumé as part of your degree program will most likely make you much more attractive to prospective employers throughout your career.

An MEng, an MBA, a Master of Engineering Management (MEM), or other degrees that help to develop core and leadership skills can open you up to a whole new world. Even if you have a business minor from your previous studies (which most engineers don't), most of these degree programs, depending on the school, will teach you about business and management depending on the specific track that you select. Why is this important? Of the many engineers (and those considering this field) who are reading this book, I would say that 50–75% will either move into a managerial position inside an engineering firm, start an engineering company, or do something other than engineering. This is based on my experience coaching hundreds of engineers in their career development.

> An MBA or engineering management degree will both give you something very important that a graduate-level engineering degree may not, which is *flexibility*. This trait is an important part of anyone's career development because it allows an engineer to pursue a wider range of opportunities.

It is true that many engineers end up in other industries. Our analytical and problem-solving background makes us very valuable to industries such as finance and consulting. Another thing to consider is that the economy at any time could force you into a career change and one of these less technically focused degrees could help tremendously in that scenario. The bottom line is that an MBA or engineering management degree may give you something very important that a graduate-level engineering degree may not, which is *flexibility*. This trait is an important part of anyone's career development because it allows you to pursue a wider range of opportunities. Remember that advancing your career is about always putting yourself in the best position to succeed.

For some engineers, a technical graduate degree may be a better fit than an MBA or an MEM degree, depending upon that engineer's goals. That is why I asked you to set your goals in Chapter 3. Make sure you are clear on where you want to go before you decide which degree to pursue.

Graduate Degree Options for Engineers (Partial List)

Master of Science in Engineering Management (MSEM)

Master of Engineering Management (MEM)

Master of Engineering in Professional Practice (MEPP)

Master of Engineering (MEng)

Master of Science (MS)

Master of Business Administration (MBA)

Master in Systems Engineering and Management (SEM)

Master of Business and Science in Engineering Management (MBS engineering management)

Master of in System Design and Management (SDM)

Master of Engineering in Engine Systems (MEES)

Master of Engineering in Sustainable Systems Engineering (SSE)

4.8   Awards Are Underrated

Here is something that is underrated or even forgotten about completely when gathering credentials: awards. Many professionals fail to highlight honors that they have won, whether it be on their résumés, company bios, or websites. You may have won some type of special recognition in college from a professional society, or one of the projects you worked on may have won an award.

In Section 4.2, I spoke about setting yourself apart from others. Another path to recognition is being the recipient of professional awards. Whether it is an employer who is considering hiring you or a client looking to assemble a design team for a project, *an award can really set you apart.* I am not saying to brag about them; just be sure to make them visible where appropriate. After all, you earned it.

Awards tell people something about you that they may otherwise not see or know, and this recognition may very well bring additional opportunities your way.

4.9   Take Advantage of Company Benefits

Many companies offer reimbursement programs for professional development activities, including reimbursement for:

- Graduate school
- The FE and PE exam fees, including books and review courses
- Fees for applications and materials for other exams such as the Leadership in Energy and Environmental Design Accredited Professional (LEED AP), among others
- Registration and renewal fees for your professional licenses and certifications
- Membership and renewal fees for professional societies
- Fees for conferences and training seminars

Please note that this list may not represent your employer's policy, as every business is different. However, these are some of the benefits that I have seen offered from various organizations. Many engineers don't take advantage of these benefits, especially when it comes to graduate school. If your employer offers financial support for graduate school and you want to pursue a master's degree, don't wait to enroll. These benefits tend to fluctuate with the state of the economy; in bad economic times, these benefits will often be reduced or eliminated.

> If your employer offers financial support for graduate school and you want to pursue a master's degree, don't wait to enroll.

4.10 Key Points to Remember

1. Credentials will not necessarily make you a great design engineer or project manager, but they will give you credibility that will make it easier to advance in your company, obtain a new client for your company, or land a job with a new employer.
2. Credentials will help set you apart from other professionals, which will assist you in standing out as an above-average engineer, greatly facilitating your advancement.
3. Strive to obtain credentials early on in your career because as time goes by your responsibilities, both personal and professional, are likely to increase dramatically, leaving you with little time to pursue credentials.
4. When you are pursuing a credential that requires an exam, prepare a study schedule by placing the date of the exam on the calendar and working backward from there. This will allow you to budget your time to cover all of the information prior to the examination date.
5. Prior to starting the application process, be sure to check your local requirements, as the rules are different in all locations and are always subject to change.
6. Submit the application for both the FE and PE exams as soon as you are eligible to do so.
7. Take a review course, if nothing else, to provide you with a regimented study schedule.
8. Remember, these exams are about getting as many questions right as you can in the allotted amount of time. Take whatever steps you can to save time, like using one reference manual and flagging the commonly used sections.
9. When pursuing a license or certification, if you do not pass the exam, immediately sign up to take the exam again the next time it is administered.
10. Consider obtaining an MBA or engineering management degree to obtain business knowledge and provide yourself with flexibility in your career. Be sure to check local requirements as in some locations an MEng may be required for licensure.
11. Be sure to note any awards that you or your company may have won. They can help you stand out (résumés, proposals, LinkedIn, etc.).
12. Be sure to take advantage of company benefit programs for professional development activities including reimbursement for exam fees and tuition for an advanced degree.

A Daily Boost from Your Professional Partner

Are you struggling to decide which credentials to obtain?
Start by looking at your career goals and figuring out which credentials will help you to achieve them the fastest.
Your goals are your road map!

Your Professional Partner,
—Anthony

References

[1] H. Irwin, Hale Irwin Quotes. Available at http://www.inspirationalstories.com/quotes/hale-irwin-you-dont-win-the-us-open-by/ (accessed on April 14, 2014).

[2] A. Bundles, Madam C. J. Walker: A Brief Biographical Essay. Available at http://www.madamcjwalker.com/bios/madam-c-j-walker/ (accessed on April 14, 2014).

[3] National Council of Examiners for Engineering and Surveying, FE Exam. Available at http://ncees.org/exams/fe-exam/ (accessed on February 3, 2014).

[4] National Council of Examiners for Engineering and Surveying, Homepage. Available at http://ncees.org/ (accessed on April 14, 2014).

[5] M. R. Lindeburg, FE Review Manual: Rapid Preparation for the Fundamentals of Engineering Exam, 3rd ed. Belmont, CA: Prof. Publ. Inc., 2010.

[6] M. R. Lindeburg, Civil Engineering Reference Manual for the PE Exam, 12th ed. Belmont, CA: Prof. Publ. Inc., 2012.

[7] M. R. Lindeburg, Mechanical Engineering Reference Manual for the PE Exam, 13th ed. Belmont, CA: Prof. Publ. Inc., 2013.

[8] M. R. Lindeburg, Engineering Unit Conversions, 4th ed. Belmont, CA: Prof. Publ. Inc., 1998.

[9] NCEES, PE Chemical Sample Questions and Solutions. Clemson, SC: NCEES, 2009.

[10] R. B. Yarbrough, Electrical Engineering Reference Manual for the PE Exam, 5th ed. Belmont, CA: Prof. Publ. Inc., 1997.

[11] M. Hansen, Mike's Civil PE Exam Guide: Morning Session. Scotts Valley, CA: CreateSpace, 2010.

[12] M. McAfee, Principles and Practice of Engineering Architectural Engineering Sample Questions and Solutions, 2nd ed. Reston, VA: ASCE, 2010.

[13] NCEES, PE Sample Questions and Solutions: Electrical and Computer Engineering (Book & CD-ROM). Clemson, SC: NCEES, 2001.

[14] NCEES, PE Civil Sample Questions and Solutions. Seneca, SC: NCEES, 2007.

[15] I. Goswami, Civil Engineering All-In-One PE Exam Guide: Breadth and Depth, 2nd ed. New York: McGraw-Hill Prof., 2012.

[16] P. Richards, Build With Steel: A Companion to the AISC Manual. Scotts Valley, CA: CreateSpace, 2012.

[17] International Code Council, 2012 International Building Code. Country Club Hills, IL: International Code Council Inc., 2011.

[18] J. Brant and G. J. Kauffman, Water Resources and Environmental Depth Reference Manual for the Civil PE Exam. Belmont, CA: Prof. Publ. Inc., 2011.

[19] R. W. Schneiter, Environmental Engineering Solved Problems. Belmont, CA: Prof. Publ. Inc., 2006.

5

Find and Become a Mentor

Mentoring is a brain to pick, an ear to listen, and a push in the right direction.

—John Crosby [1]

5.1 The Many Faces of a Mentor

The word "mentor" has been variously defined, but I think of it as a wise and trusted counselor or teacher—an influential senior sponsor or supporter. In terms of your career, I would define a mentor as a person who helps develop your professional life by assisting you in defining, pursuing, and achieving your goals and ultimately creating a successful and enjoyable career.

I use this definition because so many professionals have a goal of being successful in their career. I think this is a wonderful goal, as long as you are clear about what "success" means to you (the definition of this term is different for everyone). For me, success means being happy in all aspects of my life and having the ability to do the things I want to do, whenever I want to do them. In Chapter 3, I discussed creating your own definition of success, and there is a space in Section 3.2 where you can do so.

Engineer Your Own Success: 7 Key Elements to Creating an Extraordinary Engineering Career,
First Edition. Anthony Fasano.
© 2015 The Institute of Electrical and Electronics Engineers, Inc. Published 2015 by John Wiley & Sons, Inc.

Once you have defined success, a mentor is simply someone who will help and guide you to achieve it. A mentor will definitely support your professional development and can have a very positive influence on not only your career but also on your life. I have heard many professionals discuss the impact that mentors have had on their lives and how they helped them reach the point where they are today.

> Once you have defined success, a mentor is simply someone who will help and guide you to achieve it.

In addition to finding a mentor in your career, becoming a mentor to someone else can also be an extremely rewarding experience. Not only does becoming a mentor allow you to give back in your industry, but it also facilitates your own personal development. Mentoring can help you to develop communication skills and leadership abilities.

As you are finding out through reading this book, there are many elements to building a successful engineering career. You certainly don't want to progress through your career without any guidance. Find a mentor as early on as possible one that will help and support you in successfully reaching your goals. This chapter provides recommendations for finding the right mentor and also becoming one.

5.2 Finding a Mentoring Program and Selecting the Right Mentor

Finding the right mentor is not always easy. There are plenty of them out there, but you need to find the one that works best for you in pursuing your career goals. An important first step in finding a mentor is setting and/or reviewing your goals and understanding where you want to go in your career. Once you are comfortable with your career plan, then you can go ahead and try to find a mentor that matches up well with your vision. I recommend that you find someone who has achieved similar goals who will be able to give you specific advice on the steps you should take to get to where you want to go.

When you are ready to start looking for a mentor, you can either try to find someone directly, or you can try to enroll in a mentoring program. Should you decide to look for a formal program, start by checking with your human resources department, as many companies have such programs. If this type of program is not available through your employer, ask some of the other engineers in your firm if they have worked with a mentor or know of any engineers that might be interested.

If your coworkers can't help you, ask your supervisor for advice on finding a mentoring fit. This could be someone either within or outside the company who would be interested in mentoring a less experienced engineer. Don't be afraid to ask supervisors for help. Based on my experience, they will not think less of you for asking for assistance. Actually, your initiative and drive to better yourself will most likely impress them. You may or may not want to ask one of your supervisors to actually be your mentor. As you already have frequent interaction with this person, sometimes adding a mentoring aspect to the relationship could make your working relations somewhat overbearing. It may be best to consider someone else so you get a different perspective from what you are already learning from your supervisor.

> You may or may not want to ask one of your supervisors to actually be your mentor. It may be best to consider someone else so you get a different perspective.

If you can't find a mentor within your company, check with your local professional engineering societies. Many of them have mentoring programs that can match you with a more experienced engineer in your particular field. If you are not a member of an engineering society, I highly recommend you join one (see Chapter 7 for more on this aspect of career development). Whether or not you are a member of any professional associations, you still may be able to participate in their mentoring programs, so definitely contact them for more information.

If you still can't find a mentor through any of these sources, you can always turn to the Internet. While it would be nice to have a person specifically recommended to you, there are plenty of websites where you can connect with people that may be a good match. For example, you can use the social networking site LinkedIn to search by zip code for mentors in your area (I discuss LinkedIn in detail in Chapters 1 and 7, including using the Groups feature to connect with other industry professionals). There are also online engineering communities like engineeringcareerdevelopment.com that will give you access to engineers all over the world. Depending on your comfort level with the different forms of communication, the mentoring relationship can work in person, over the telephone, or through web applications (i.e., Skype), so don't put geographic limits on your search.

5.2.1 Try to Select Someone from Your Specific Discipline

Whether you enroll in a formal mentoring program or seek individual referrals from colleagues, you may have to select a mentor from a pool of multiple candidates. The following are some recommendations for selecting the right mentor for you.

For example, civil engineering is a very broad engineering category that is filled with many different disciplines. Civil engineering includes, among other subjects, structural engineering, geotechnical engineering, transportation engineering, environmental engineering, etc. If your career is focused in the environmental engineering realm, it would be best to select another environmental engineer as your mentor, even though a transportation engineer is also considered a civil engineer. Selecting a mentor closely related to your own specialization is a benefit because (as in this example) no one will know how to advance in the environmental engineering field better than an environmental engineer.

In addition to selecting someone in your specific discipline, I would also recommend you select a mentor who has achieved goals similar to those that you are pursuing. The only person better to help you than an experienced engineer in your discipline is one that has already achieved the goals that you are striving for. Why reinvent the wheel and try to figure out from scratch how to achieve a goal when someone who has already achieved it can share with you the steps? Worksheet 11-6 in Chapter 11 of the book will help you to identify the right mentor for you. The worksheet also includes a template e-mail that you can use to reach out to a prospective mentor.

> Select a mentor who has achieved goals similar to those that you are pursuing. The only person better to help you than an experienced engineer in your discipline is one that has already achieved the goals that you are striving for.

5.2.2 Consider Your Level of Comfort

You should also consider how comfortable you might be discussing personal and confidential issues with this person. (This is another reason that it might not be a good idea to select a mentor within your own company.) You may look to your mentor to provide advice on salary negotiation or conflicts with staff in the workplace, which may be awkward to discuss with a fellow employee or supervisor.

5.2.3 Don't Settle on the First One That Comes Along

Last, when it comes to selecting a mentor, do not settle on the first person that volunteers. A mentor is critical to your career and personal development; approach the selection process with this in the forefront of your mind at all times. It's imperative that you find the right person to help you define and achieve your goals. Failure to do so can really stall your engineering career. While there may be temptations to work with the first person that offers, be sure to follow the recommendations in this section and find the best match for you.

In the end, it's important to know that it may take some time and effort to find the right mentor for you, but if you do your research and select the right person, your mentor can have a profoundly positive impact on your career. If you don't find the right person the first time, try again. There are so many people out there that want to help you—you just have to ask.

5.3 The Mentoring Relationship for Protégés

The mentoring process is analogous to many of the computer programs that you use as an engineer; what you get out of it is directly related to what you put into it. Unless you are part of a formal mentoring program, it will be up to you and your mentor to set the guidelines of the relationship. Establishing these expectations up front is critical to the success of your mentoring relationship.

Many professionals say that they have mentors however, they only meet or call them a few times a year. In my opinion, that is not a true mentor—it's more like chatting occasionally with an old friend. In a solid mentoring relationship, the mentor and protégé meet once a month at a minimum. I recommend meeting weekly or every other week for the first few months. These meetings don't have to be long, but when you meet more frequently, it helps to keep the momentum going and gives the mentor more opportunities to help you take action. When too much time elapses between meetings, you tend to let other priorities become more important, and your career advancement efforts may take a backseat.

5.3.1   Establish Levels of Confidentiality

The most important aspect of a mentoring relationship is confidentiality. This relationship needs to be a supportive arena where you as the protégé feel free to openly discuss whatever is affecting your career development. The topic of confidentiality should be discussed at the beginning when you both are creating the guidelines for the relationship. If confidentiality is not discussed and agreed upon, then you may not feel comfortable opening up to your mentor about some of the biggest obstacles you are currently facing. This is another reason that it may be more beneficial to work with someone outside of your company, as you will most likely feel more comfortable telling that person exactly what is happening in your career. Your mentor should be very agreeable to the confidentiality component of the relationship, especially if he or she wants to be sure that the relationship benefits you as much as possible.

> The most important aspect of a mentoring relationship is confidentiality. This relationship needs to be a supportive arena where you as the protégé feel free to openly discuss whatever is affecting your career development.

While confidentiality is an important aspect of the mentoring relationship, even more important is starting the relationship off in a positive way and setting yourself up for success. To do that, it's best to spend the first few sessions with you mentor getting to know each other. You can explain or set your career goals with his or her assistance. Establishing your goals with a mentor will provide a solid baseline for the rest of your time spent together. Once you have laid out clear goals, it will be very easy for you and your mentor to prepare and execute a plan and track your progress regularly.

Sample Goals You Might Set with Your Mentor

- Develop a study plan for taking and passing the Principles and Practice of Engineering (PE) exam.
- Create an outline with the steps I need to take to help me reach the project manager I level (or any other position you wish to obtain in your company).
- Join and get actively involved in a professional society.

5.3.2   Set Expectations for Mutual Accountability

Another important aspect of the mentoring relationship is accountability. Depending on the frequency of your mentoring sessions, consider creating action steps or setting goals during each meeting, which you plan on achieving before the next meeting. This will help you to become action oriented in your career advancement efforts and establish a self-imposed deadline for goals. In my experience with mentoring and even in my work as an executive coach, I find that accountability is one of the key elements in helping professionals take their career from where they are now to where they want to be going forward (see the next section).

> Remember that when it comes to your mentoring relationship, what you put in is what you get out.

Always keep in mind that your mentor is volunteering to help you, so be sure to always be respectful of his or her time. If you can't make it in one of your meetings, try to give your mentor plenty of notice so that schedules can be adjusted. *Remember that when it comes to your mentoring relationship, what you put in is what you get out.* From the very beginning, be sure to establish clear guidelines for your relationship that include scheduling, confidentiality, and accountability and also discuss your career goals. To cement these guidelines, it might help to write them down and share a copy with your mentor that you can refer to regularly. If you follow these recommendations, you will be setting yourself up to create a very rewarding relationship, for both you and your mentor. Included in Chapter 11 of this book, Worksheet 11-7 is a checklist to make it easy for you to follow the guidelines laid out here throughout the mentoring process.

5.4 The Importance of Accountability

How many times do you say you want to do something in your career, but then (for whatever reason) you don't do it? Most people do this multiple times in a week. Accountability can help carry your ideas through to actions; and in a mentoring relationship, a mentor can help to provide the structure to keep you motivated and inspired.

Do you remember when you were in grade school? You would come home in the afternoon and wanted to go outside and play with your friends or tackle your favorite video game; but, instead, you sat down and did your homework. Why did you do that? Most children would tell you that they did it because their parents made them do it. I doubt their parents physically made them do it; however, their presence, as well as previous disciplinary action, helped to hold the child accountable for assigned responsibilities. The choice was either do the homework or suffer the consequences of receiving bad grades.

A mentor provides the same accountability. Your mentor will not come into your office and help you sign up for an exam or apply for a new degree, but you know that you will be asked at your next meeting why you didn't follow through. That's the power of accountability.

If you do not currently have a mentor, you can hold yourself accountable by discussing your career goals with your supervisor. By doing this, you are making a commitment and are instilling accountability within yourself, because now you know that your boss will be waiting to see if you reached your goals or if you are taking action steps toward achieving them. This works great with signing up for exams or certifications. Once you tell your supervisor that you will be sitting for the next PE exam, you can bet you will start studying, and soon. This can also be accomplished through documentation that you prepare for annual review, which I will discuss in detail in Chapter 8.

> Always try to hold yourself accountable in your career. When you do, your performance will be at its peak.

5.5 Getting the Most from Your Mentor

Once you have established guidelines and you start meeting with your mentor, it's important to be prepared for your meetings so that you get the most out of the relationship. It would be helpful to prepare a list of questions before each session that you can ask during the meeting. I recommend creating a document on your computer or keeping a notepad on your desk so you can jot the questions down as you think of them during the day.

Ask your mentor questions such as "What were some of the biggest challenges that you faced in your career and how did you overcome them?" You will find that many of the obstacles he or she overcame are the same that you face, too. Even if the challenges may not be the same, the strategies used to overcome your mentor's hurdles may be helpful to you in your situation.

Try to ask open-ended questions that can't be answered with a simple yes or no, as you want to get as much useful information as possible during these meetings. If you are pursuing certain credentials or preparing for exams like the Principles and Practice of Engineering, ask your mentor for advice in these areas.

Preparing for your meetings in advance is critical to the success of the mentoring relationship. If you are too busy to prepare for a meeting, ask your mentor if you can postpone it until you do have time to organize your thoughts and questions. It is not always easy to find a mentor, so once you have one, take full advantage of every minute you have to spend with this experienced professional who has committed to helping you achieve your goals.

5.6 Become a Mentor

While having a mentor is an important part of advancing your career, being one is just as important. Giving to others will help you grow tremendously, both personally and professionally.

If you are wondering about how you are going to find your mentor, I urge you to also think about how you are going to become one. You can become a mentor at any age as there is always someone who can benefit from your knowledge and experience. Whether you are a retired engineer mentoring younger engineers or a younger engineer mentoring college students preparing to become engineers, you can have a positive impact on others by reaching out and offering your assistance. Several members of my Institute for

Engineering Career Development have started mentoring programs at their alma maters utilizing the information in this chapter, and you can too.

> One of the engineers that read the first edition of this book was so inspired that he created and implemented a mentoring program at Rensselaer Polytechnic Institute (RPI). What might you accomplish with a little inspiration?

As you help others, you will see new skills of your own developing that you most likely didn't even know you had. Mentoring will help you develop communication skills and leadership abilities. As you start giving back to others while building your career, you will feel strongly inclined to continually give more. In fact, that is how this book came about. Once I started coaching, training, and speaking to other engineers about their careers, I felt it was my responsibility to share my knowledge with not only those I have worked with individually but also engineers all over the world. There is no greater reward than the feeling you get when you are able to help others.

So, if you think you're too young or too old to be a mentor, think again. If you assume you don't have anything valuable to share with people, please reconsider that position. There are engineers out there who will benefit greatly from what you have to offer, and they are just waiting for your call.

5.7 Selecting the Right Protégé

Once you have decided to become a mentor, the obvious next step will be to find a protégé to work with. The best way to go about doing this is to reach out to your human resources department should you work in a larger engineering company and/or the local engineering professional associations. Express your interest in becoming a mentor. You will most likely be enrolled in a structured mentoring program that will match you with a less experienced engineer.

You shouldn't just agree to mentor the first engineer that they attempt to match you with. Even though you are volunteering your time, you owe it to yourself and your prospective protégé to ensure that you are a good match for each other and that you can actually help. Let's face it—regardless of how much experience you have as an engineer, you won't be able to help every younger engineer.

Attempt to work with a protégé who has similar goals to the ones that you have achieved in your career. For example, if you were a licensed structural engineer working on high-rise buildings in California, try to find a younger engineer with those same aspirations. Even if he or she is not working specifically in California, try finding a structural engineer who is looking to get licensed and work in the field of high-rise buildings in earthquake-ridden areas. You are much better equipped to help an engineer on this career path rather than a civil engineer in the wastewater or transportation fields. It is also usually more enjoyable when you can share all of the intricate details of your career story, which is more doable if the protégé is in your exact field or one closely related.

Do not be afraid to turn down a protégé if you don't feel like he or she is a good match for you. While you might feel bad to do so initially, in the long run, you are saving you and your protégé valuable time that can be used in other areas of career and personal development.

5.8   Being the Best Mentor You Can Be

Once you find the right protégé, the mentoring relationship may begin. Remember, confidentiality and accountability are key factors in a successful relationship and should be discussed in your first session. It is also important that, as the mentor, you provide some guidelines for how the actual relationship will work; begin with setting the time of your meetings. You and your protégé should agree to meeting at a similar day each week or month to ensure that you do meet on a regular basis. Without consistent meetings, it is very difficult to make progress in any initiative, specifically career development. We all know that schedules are subject to change, but at least schedule something, like lunch on the first Friday of each month, and then reschedule if needed later on. Usually, when you add an appointment to your calendar early enough, you make it a point to keep it and work around it.

During your first meeting, I highly recommend that you both spend some time outlining goals for your work together. Ask your protégé a series of questions related to where that person is career-wise and where he or she would like to be in six months to a year from now. From those answers, help your protégé to formulate clear, concise, achievable goals for that time period. It is important that these goals are achievable, because if they are too lofty to begin with, your protégé may see them as unreachable or difficult and may stop pursuing them before starting. To really foster success, give very minor tasks for the first few weeks to help them gain momentum. For example, if your protégé wants to obtain a PE license, create a plan for the following meeting (i.e., the assignment is to download the application and read it for discussion with you). This is a simple task, yet it will forward action and put a heavy emphasis on action in the relationship.

Meeting with your protégé in person from time to time can be extremely helpful in facilitating a successful relationship. While there is nothing wrong with having your mentoring meetings over the phone, if possible, the in-face meetings can add another dimension to your relationship and encourage a deeper connection, which will result in more progress. I provide career coaching to engineers on a regular basis by phone, and whenever I meet someone in person, I have found that person to be more comfortable with subsequent calls and the atmosphere is much more relaxed. Again, this is not a "must," but it is highly recommended if possible.

As a mentor, it is important to remember that your job is not to tell the protégé how to develop a career. Your protégé should not be coming to you for specific instructions on how to succeed (and if he or she is, you need to redefine the working relationship). Your job is to guide that person in his or her career development efforts toward achieving outlined goals. Listen to individual needs and questions before you start to provide recommendations. Don't assume a protégé needs one thing, because the need may be something completely different. Take a listen-and-respond approach rather than tell, tell, tell. Of course, if your protégé gets off track, provide gentle guidance. This approach will not only ensure that the protégé is staying aligned with personal goals, but it will also provide a comfortable atmosphere for open conversations.

> "We're here for a reason. I believe a bit of the reason is to throw little torches out to lead people through the dark." —Whoopi Goldberg

Last, but certainly not least, strive to always be a resource for your protégé. Constantly remind your protégé that help is readily available. This often makes people feel much more comfortable reaching out. Also, if you happen to see things like articles, seminars, or books that might be helpful to your protégé, share these items. Your protégé should consider you a guide, a friend, and most importantly a resource for those much-desired career goals. Worksheet 11-8 in Chapter 11 of this book provides a checklist and some help in this regard.

5.9   How to Graciously End a Mentoring Relationship

At some point in time, for both a protégé and a mentor, it will be time to end the relationship. A protégé will get to a point where a mentor has helped as much as possible and it may be time to find a new mentor. On the other hand, a mentor may feel that he or she can no longer help the protégé or volunteer any more time in the relationship. In either instance, this section provides some recommendations for delicately ending the mentoring relationship.

When ending relationships with your mentor or protégé, it is important to be completely honest with the other person, providing a meaningful explanation as to the shift of the relationship. If the relationship is no longer helpful to you as a protégé, tell the mentor tactfully, keeping in mind constructive criticism. As a mentor, if you feel that you can't help the protégé any longer, let the discussion reveal that you have taken it as far as you can, perhaps suggesting another mentor who can provide additional help. This will help to prevent the protégé from losing confidence and possibly feeling personally at fault.

Always thank the other person for the time when ending a mentoring relationship. Even as the mentor, it is courteous and respectful to thank the other person for the attention and the time spent together. This will allow both parties to walk away feeling good about the relationship. Maintaining a positive tone and energy throughout this conversation will make it easier to leave on good terms. Just because you are ending the mentoring relationship, doesn't mean you can't stay in touch and maintain a positive, fruitful friendship.

Finally, please be sure that you do end the mentoring relationship when the time is right for you. Participating in a relationship like this when you are not committed to it fully is only wasting time on both sides. The mentoring relationship can be a rewarding one, but like all things, it, too, will come to an end. Following the recommendations in this section will help ensure that the end happens at the right time in a positive way.

5.10 Actions to Avoid for Mentors and Protégés

I always try to maintain a very positive attitude and mental approach in everything I do; however, in this section, I think it is more helpful to list some of the things not to do for both the protégé and mentor:

Protégés don't:

- Miss appointments
- Show up to a mentoring session unprepared
- Select a mentor that has no familiarity with target industry or goals
- Limit themselves to one mentor
- Think that they can go it alone

Mentors don't:

- Miss appointments
- Push protégés in a direction they don't want to go
- Just give answers (but they do ask questions)
- Forget that they learn, too, from the relationship

5.11 Key Points to Remember

1. A mentor is a person who helps you to develop professionally by assisting you in defining, pursuing, and achieving goals and ultimately creating a successful and enjoyable career. *Find a mentor as early on as possible in your career.*
2. To find a mentor, check with your human resources department, coworkers, or supervisors to see if they can recommend a program or a specific person. If you can't find a mentor within your company, check with local professional engineering societies. If you still can't find a mentor through any of these sources, use the Internet to find online engineering communities through which mentors may be available.
3. Confidentiality and accountability are extremely important aspects of the mentoring relationship. Be sure they are cemented as part of your guidelines from the beginning to maximize the success of the relationship.
4. If you do not currently have a mentor, hold yourself accountable by discussing your goals with your supervisor. This will help you follow through on your commitments by knowing that you will have to answer to your supervisor if you don't.
5. To get the most out of your mentoring relationship, prepare in advance for each meeting. Try to prepare open-ended questions that you can ask your mentor during the meeting to gain as much information as you can in the allotted time.

6. Consider becoming a mentor, regardless of your age. Through helping others, you will see new skills of your own developing that you most likely didn't even know you had. Mentoring will help you to develop your communication skills as well as your leadership abilities.

7. By selecting a protégé in your field, you will create an atmosphere that will allow for you and your protégé to get the most out of the relationship.

8. As a mentor, be sure to help your protégé set achievable goals at the end of each meeting to keep him or her consistently moving forward.

9. Whether you are a mentor or a protégé, be sure to end the mentoring relationship when the time is right for you. Be honest, courteous, and positive when ending the relationship so both parties walk away feeling a sense of accomplishment.

A Boost from Your Professional Partner

As much as you think you know about your field, someone else knows more. Find a mentor. Everything will get easier if you do this.

Your Engineering Career Coach,
—Anthony

Reference

[1] Arizona State University Provost Office, Mentoring (Crosby quote). Available at https://provost.asu.edu/academic_personnel/mentoring (accessed on April 30, 2014).

6

Become a Great Communicator

The single biggest problem in communication is the illusion that it has taken place.
—George Bernard Shaw [1]

6.1   In Today's World, Communication Is a Whole Different Ball Game

Long gone are the days when the primary way of communicating with someone was to pick up the phone and call. Good or bad, technology has provided us with many different ways to communicate, such as e-mail, text messaging, web cameras, blogs, social networks, instant messaging, and more. You would think that all of these additional tools would make it easier to communicate, but that is not always the case. At times, having all of these different options can make it difficult to identify the one method or channel that is most efficient for the situation at hand.

As an engineering professional, you will most likely spend a good portion of your career working in a teamwork environment. Whether it is a group of your coworkers or outside consultants, your ability to communicate with them will directly impact your career success. Taking it one step further, you will also be challenged with communicating technical information to nontechnical people. (I don't know

Engineer Your Own Success: 7 Key Elements to Creating an Extraordinary Engineering Career,
First Edition. Anthony Fasano.

about you, but I cannot remember there being a required course in school for handling that in my degree program, but every engineering major should require one.)

You may be an excellent technical professional who can crunch numbers, prepare amazing graphs, and analyze results with the best of them; however, if you can't communicate those results to others in a powerful way, then what is the point of producing them? If you cannot communicate your work, then your work does not exist to the outside world. And if you cannot communicate it with power and grace, then you may do some damage to your credibility.

> You may be an excellent technical professional who can crunch numbers, prepare amazing graphs, and analyze results with the best of them; however, if you can't communicate those results to others in a powerful way, then what is the point of producing them?

This chapter will provide strategies for improving your communication skills and increasing your confidence to help you become an effective communicator and converse clearly in any situation that you encounter.

6.2 Project/Team Communication Starts In House

In some respects, technology has made it more of a challenge to communicate with your colleagues (coworkers as well as other consultants) or at least to keep the communication consistent.

The key to communicating in today's world is to make sure that everyone is on the same page. Your means of communicating, whether it is by phone, e-mail, or any other type of web application, is not the most relevant factor. What's most important is that you keep everybody up to date on the status of your project, as failure to do so can negatively impact the success of the project, the cohesiveness of your team, and, most importantly, your client relationships. For example, how do you think one of your clients would feel if you called to ask the same question that another one of your team members had just asked 30 minutes ago via e-mail? Your client is going to feel like time is being wasted, and it will be immediately apparent that your team is not coordinated in its communication efforts.

To avoid any type of communication mix-up, you must have some kind of system in place that makes it easy to keep everyone up to date in real time. Perhaps, institute a team or company policy that says before someone reaches out to a client, the person sends an e-mail to the *entire* team to keep them in the loop. At a minimum, each team member should be copied on the correspondence if it's being sent over e-mail. Another option may be to have a project website where everyone can see exactly what's going on, again in real time. This may sound cumbersome, but with today's technology, anything is possible and usually with minimal effort.

Having a good communication system in place could mean the difference between delivering an average project or a great one. More importantly, good communication can determine whether or not your relationship with your client will be long lasting or short-lived, and we all know the value of repeat clients.

> Good communication can determine whether or not your relationship with your client will be long lasting or short-lived, and we all know the value of repeat clients.

Be sure that when you work with a team, you establish these communication guidelines as early as possible to avoid conflict with the team. When people don't communicate openly, they tend to get upset with each other, which can, in turn, affect the quality of their work and put a damper not only on the project but also on your career as a whole. When people don't communicate with each other, they start to make assumptions about what the other person is thinking or what actions they are going to take. These assumptions may lead to decisions that negatively impact the team, the project, and the company as a whole.

Remember that communication within your team is about keeping everyone informed and up to date on what is going on. This is not necessarily an easy task, but it is one of the most important ones, considering the impact that teamwork will have in meeting the client's needs. If you can't ensure a satisfied and happy client, believe me, your career will be anything but successful.

6.3 Communicate Early and Often

Whether you are preparing a design for a new project or overseeing the construction of an existing one, it is important to communicate with the clients as well as the end users of whatever you are designing or constructing as early on as possible. For example, if you were designing a new park for a community, wouldn't it be helpful to find out from the citizens in the community what their vision of that park is? Not only will it make your job as a designer easier, it will also ensure that you are meeting the needs of the community members who will then be on your side.

One of the reasons that it is so important to get the community on your side is that you will find that if you do things without asking people about their needs, the result may be that they end up speaking out against your project. Continuing with the park example, imagine that you went ahead and designed it without soliciting any input from the people who are going to use it; this could cause them to speak out negatively against the project and delay the entire approval process, making your job much more difficult than it should be.

The bottom line is that the end users have every right to offer input on a project because they will be the primary users for years to come. Be sure you remember this fact because many engineering professionals just want to design, design, design, and they forget the impact that their design will have on communities. This is one of the beautiful things about being part of the engineering industry—you have the ability to impact thousands or even millions of people on a daily basis through your work as an engineer.

> You have the ability to impact thousands or even millions of people on a daily basis through your work as an engineer.

The same sense of responsibility applies to your relationship with your client. Even if the client isn't necessarily the end user, there still may be certain requirements for how the end product should be, and it is up to you to ask for and listen to the client's ideas early on in the process to avoid conflict and additional work later on.

Communicate early and often. Not only will it help you to save time and money, but it also will help you to serve citizens, clients, and your community in the best way possible.

6.4 How to (Almost) Explain Rocket Science to a Nontechnical Person

One of the challenges that you will be faced with in your career as an engineering professional is that of trying to relay technical information to someone with a nontechnical background. It might be a client, someone in the legal department, someone outside of your field, someone at a conference, or a supplier. No matter who your nontechnical audience is, this is one aspect of your career that if done properly can really set you apart from other engineering professionals, because many of them cannot do this well.

To master the art of relaying technical information to others (and, yes, it is an art), you must start by thinking about the information you are presenting from the other person's perspective. When doing this, put yourself in the other person's shoes and think about what would help them to easily understand the information.

> "Empathetic people are superb at recognizing and meeting the needs of clients, customers, or subordinates. They seem approachable, wanting to hear what people have to say. They listen carefully, picking up on what people are truly concerned about, and respond on the mark." —Daniel Goleman, Primal Leadership [2]

It can also be very helpful to use analogies to put your information into terms that your audience will understand. Everyone has a unique perspective based on individual background and experience, and by using terms that are familiar, you can make others much more comfortable and receptive to your message. If you cannot find a good analogy or example that relates to those you are speaking to, you can always revert to the financial aspect of that situation, assuming it is relevant. For example, if you are planning to change the design on a project to make it more cost effective for the client, then you probably won't need to get too deep into the technical jargon. You can simply relay the fact that money will be saved with this new design and, of course, give the dollar value, which the client will certainly understand.

The ability to communicate clearly is a critical engineering skill because it allows you to convey to clients how valuable you are to them. If they cannot see the value you are providing, then you are no different than the next engineer that can prepare a design or do an inspection. I have worked on projects where we had to squeeze unbelievable amounts of stormwater treatment onto a very small site. In situations like this, the client is never going to appreciate the amount of work or skill that goes into making the site work by simply looking at a plan. That is when you must utilize some of the strategies given in this chapter to explain or attach a value to your work. Then, when your client is assembling a design team for the next project, you can bet that you will be one of the first called because they know how important you and your company are to their success.

> Communicating clearly is a critical engineering skill. Only good communication can impart value to your client. If a client can't see the value you are providing, then you are no different than any other engineer who can prepare a design or do an inspection.

I am not saying that you have to walk around all day long bragging to your clients about how great you are. What I am saying is that in this competitive corporate world that we live in, *those individuals and organizations that can clearly communicate their value to their clients will be the ones who will always have projects.*

6.5 Honesty Really Is the Best Policy

Because of the nature of the engineering industry, you will communicate with many different people on a day-to-day basis. If there is one hard and fast rule that you need to set for yourself when communicating with people, it should be to always be honest. The old saying "Honesty is the best policy" is definitely true.

As an engineering professional, you will face many project deadlines in your career, and you will be presented with many opportunities where the pressure may push you to "bend" the truth or stretch the boundaries of ethical behavior. This may be even more common in the earlier stages of your career where you are trying to make a good impression on your coworkers, supervisors, and clients. Younger engineers often find themselves telling others exactly what they want to hear whether it is true or not, just to be sure that their reputation doesn't get tarnished early on in their career. While you might feel that lying may be convenient in some cases, whether it be in your career or in your personal life, it always catches up with you.

I remember a specific case early on in my career where I had been assigned a new project to manage. It was one of the first projects in which I was going to be managing everything, including the relationship with the client. This was a new client for the company, and I was very excited to show the client what we were capable of producing. One day, the client called me and asked if we could have a certain plan ready for him the next day. I knew there was no way that we could have it done in that time frame, but I didn't want to let this client down. So, I told him that we would have the plan ready the next day. I told him exactly what I knew he wanted to hear. At that moment, it was the easiest way to respond to the client's request. My client was pleased with my response and thanked me graciously. However, he didn't sound so gracious the next day when he was advised that his plan was not ready. It was a very uncomfortable situation and left the client with a negative impression of my employer and me.

From that point on, I decided to never lie to any of my clients again, or anyone else for that matter, even if I thought it would help me get ahead in my career. After that day, if a client asked me for something in a certain period of time and we couldn't have it ready that quickly, I was totally honest with my response. I would say, "I know this is something you need urgently and we will start working on it right away, but I just want to make you aware that it usually takes us two to three days to do something like this." I learned how to tell them the truth in an honest but reassuring way. In fact, by taking this approach, I ended up eventually developing a very good relationship with the client I discussed above that I had previously let down. At times, he would be upset with me in the moment if I didn't tell him what he wanted to hear, yet in the long run, he developed a tremendous amount of respect for me because he knew he could always count on me to tell him the truth, good or bad.

Being honest also applies to your supervisors and everyone else, for that matter. I know of many engineers who do not like the specific job that they are doing, but they are afraid to tell their supervisor for fear of losing their job. In most cases, a supervisor will appreciate your openness and honesty and will try to accommodate moving you into a position that is more enjoyable for you. It benefits both you and the company to do so. If there are no other positions available at that time, at least you have made management aware of your feelings, so if something comes up, an offer may be given to you before they hire someone else for the position. I know there is an outside chance that a move like this could cause you to lose your job; however, failure to speak up in these situations can lead to years of unfulfilling work and a purposeless career. In my opinion, that is a greater pain than losing an unfulfilling job.

> "Honesty and integrity are absolutely essential for success in life—all areas of life. The really good news is that anyone can develop both honesty and integrity."
> —Zig Ziglar   [3]

I worked with an engineer once who lied to everyone about everything. It got to the point where the clients would call me, the younger engineer, rather than him as the project manager, because they knew that they couldn't trust a thing that he said. To this day, people avoid him because of the negative reputation that his dishonesty has created for him, and believe me, this is something that is not easily shaken.

If you find yourself always telling people what they want to hear rather than telling them the truth, please consider taking a different approach and be totally honest with them. You will be happy you did as again, I repeat, honesty really is the best policy.

Honesty and integrity are absolutely essential for success in life—all areas of life. The really good news is that anyone can develop both honesty and integrity.

6.6 How You Say Something Is Just as Important as What You Say

It amazes me to hear how people talk to other people throughout the course of the day. People don't realize that the way they say something is just as important as what they say. When I make this statement, I am referring to the exact words you choose, the tone you use when you say them, and the speed at which you speak.

When you talk to people, be sure that you do so in a way that will both appeal to them and help you get what you want out of the conversation. For example, let's say an engineering manager needs an administrative assistant to fill out some forms for an application on a project, saying "I need you to get this application done right away so that we can send it to the client for a signature." How do you think that assistant is going to respond to that request? Starting the sentence by telling someone to do something right away will most likely be upsetting. The talks may still get the job done, but a coworker certainly won't appreciate the tone. Continued use of this type of language can strain any relationship and ultimately affect the success of the team.

Instead, frame requests with respect. Perhaps, the manager could say, "I have a very important application that I would like to ask you to complete for this project. Unfortunately, it has to be done fairly quickly. Would you be able to fit this into your schedule today?" Do you see the difference? In this case, the manager acknowledges the assistant's certain area of expertise and is being courteous by asking if the task can fit into the schedule. The rephrased approach demonstrates respect of the time and talent of the coworker, not just barking orders.

Those people that do not understand the place of respect at work often ask me, "Why should I say it so nicely? I am only asking them to do their job." Be careful of this mentality. Yes, it may be their job, but you should be grateful for what they are doing and reflect that through your actions.

> "Really big people are, above everything else, courteous, considerate and generous—not just to some people in some circumstances—but to everyone all the time." —Thomas J. Watson Sr. [4]

There are several reasons why it is important to be polite to people when speaking to coworkers, especially when asking them to do something for you. First of all, it is simply good manners to be polite, and you will be much more likeable if you are. Remember that in order to have an extraordinary career, you are going to want to always put yourself in a position to succeed. Do you think you are in a better position if people like you versus disliking you? I don't think I have to give you the answer to that question. You want to earn the respect of your colleagues, and being polite is one way that you can do this.

One approach you can take in being more courteous and respectful is asking questions of people, which will allow you to get what is called a "buy-in." In the example I gave earlier where the engineering manager asked, "Would you be able to fit that into your schedule today?" the engineering manager was politely asking the administrative assistant to do something. Likely, the assistant is not going to say no, because like many of us, that person probably has a natural desire to succeed. However, by saying yes, the assistant is committing to completing the task. The manager didn't tell the assistant that it had to be done that day, but it might be implied. Think about it. If your boss asked you if you could complete a task before you left for the day and you said yes, you would stay at work as late as you had to in order to get it done because you said you would. However, you might be upset with your boss, and that resentment might carry forward and impact the team's success in the future.

The tone of your voice can also have an effect on how your words are received. You don't want to use a tone that sounds condescending. You want to use a pleasant tone so that people enjoy conversing with you. The volume of your voice is also important. When you speak too loud, it can turn people off, while speaking too softly can make it difficult to get your point across. You have to develop a tone and volume that are comfortable for both you and those that you converse with.

> The tone of your voice can have a big effect on how your words are received. You don't want to use a tone that sounds condescending, ever.

Finally, the rate at which you speak can dramatically impact the quality of your conversations. I'm from New York City, where we all move at a super fast pace, and I have always struggled with trying to speak slowly when I travel outside of the metropolitan area or when I give a presentation. Speaking too fast can cause you to forget important words that can alter the message you are trying to convey. However, speaking too slow can make it difficult for people to follow you. Again, this is something you will need to work on until you achieve a speed that is comfortable for yourself and others. In the next section on public speaking, I will give you some strategies for working on your tone, volume, and rate of speech.

I challenge you to start being more aware of the way you communicate. Improving your choice of words, tone, and volume may take time, but even a small change can have a dramatic impact.

6.7 Public Speaking: The Ultimate Differentiator

If you were to choose only one thing to do to improve your communication skills, I would recommend developing your public speaking skills. Many engineering professionals, especially those early in their career, don't realize how important it is to be a good public speaker. The good news is that the ability to be a good public speaker is not one of those "you either have it or you don't" skills. It is something that you can learn and develop over time (Table 6.1).

TABLE 6.1 Public speaking statistics

Public speaking statistics	Data
Percent of people who suffer from speech anxiety	74%
Percent of women who suffer from speech anxiety	75%
Percent of men who suffer from speech anxiety	73%
Number of Americans who have a social phobia	5.3 million
Number of Americans who have a fear of crowded or public places	3.2 million

This table shows the percentage of people who suffer from fear of public speaking [5].

As an engineering professional, your ability to present to people, whether it is two or 2000 people, can make or break your career. I realized this early on in my career, yet I didn't take the time to develop this skill until later on. Once I realized that I had powerful information and strategies regarding career advancement that I wanted to communicate to professionals, I decided that I was going to learn how to speak effectively so I could get my message out to the world.

"Speech is power: speech is to persuade, to convert, to compel." —Ralph Waldo Emerson [6]

To develop my speaking skills, I joined a local chapter of Toastmasters International. Toastmasters is a nonprofit educational organization that operates around the world for the purpose of helping members improve their communication, public speaking, and leadership skills. There is a good chance that there is a chapter near you. Groups like Toastmasters help you to not only develop speaking skills, but they also help participants to increase their confidence, which helps to become a great communicator. In these groups, you will learn how to speak spontaneously, and you will also be given the opportunity to prepare and present longer speeches with specific goals and objectives. One of the nice things about these groups is that they provide such a comfortable, supportive atmosphere for you to practice your public speaking. Your audience is comprised of people just like you who want to learn how to develop their skills. Odds are high that every person in the group was just as fearful as you may be, if not more, of public speaking at some point.

Having great public speaking skills is an absolute MUST in order to have an extraordinary engineering career. I would say that most engineering professionals are not very good public speakers, but that doesn't mean they can't be. Unfortunately, engineers don't see public speaking as an important aspect of their career development, and therefore, they don't take the time to develop this skill.

I cannot emphasize enough how many different ways public speaking can impact your career in a positive manner. In addition to improving your communication skills, public speaking will help you to bring new business into your firm. Prospective clients may see you presenting a project at a local board meeting; a good talk may be all they need to call your firm next time a project comes up. Your company may be impressed enough with your presentation skills that they ask you to accompany a marketing professional in meetings with prospective clients. These skills will also prepare you for being a manager, if you are not one already. Since managers have to be able to clearly communicate messages to their teams, your employer may take your public speaking skills into account when considering you for a promotion. Improving your public speaking skills also helps you to build your confidence, which is something that many professionals don't focus on enough. Practice also helps you to develop the tone and volume of your voice, which we discussed the importance of in Section 6.6.

> Outrageously successful engineers are great presenters who not only get their point across but do it in a clear, concise, and confident way. They excite and involve their audience.

After giving one of my Engineer Your Own Success seminars a few years ago, an engineer, in his sixties, approached me and said that he wished someone had imparted upon him the importance of public speaking earlier on in his career. To this day, he is terrified to get up in front of a room, and because of his fear, he has always been at a desk in a very technical role and never attempted to get on the track to management, which he told me he regrets to this day. This is an example of the profound impact the ability to speak in public can have on your career and your life.

If you are reading this and are terrified of public speaking, I have great news for you. With some hard work, determination, and practice, you can become a great public speaker. Many engineers are good presenters and are effective in getting their point across so that people understand them. However, outrageously successful engineers are great presenters who not only get their point across but do it in a clear, concise, and confident way while exciting and involving their audience. This is your opportunity to take a part of your career that may have been lacking and transform it. Join a group, read a book, or watch some videos online about public speaking and start making improvements today. This skill can be the ultimate differentiator in your career development.

6.8 How to Improve Your Public Speaking Skills

In the last section, I emphasized how important your public speaking skills are as an engineer and recommended joining Toastmasters to start developing yours. Now, I am going to give you some specific strategies for improving your public speaking skills. Depending on your discipline and specific job description, you very well may have to deliver multiple presentations per week throughout your engineering career. The quality of your presentations will undoubtedly have a huge impact on your engineering career development, specifically on your ability to advance from engineer to manager.

As I mentioned earlier, public speaking is not a talent or gift that you are born with—it is a skill that must be developed and constantly improved. The best public speakers in the world were probably just as scared as you are to get up on stage at one point in their lives. That is not just good news; it is great news, because it means that no matter how much you fear public speaking, you can overcome your fear and shine with the best of them.

The following are some key pointers for developing the speaker inside of you:

1. *The more you speak in public, the better you get.* This is the cardinal rule in improving your public speaking skills. You must get up on stage in front of real live people and present. Please don't stand in front of your mirror talking to yourself—it doesn't work. The difference between standing in front of a mirror and standing in front of a crowd is monumental. You can't capture that feeling of looking out over 20 or 2000 people. You have to experience it several times to be comfortable with it. Every time you do it, you become more and more comfortable in front of the crowd. To accomplish this task in a safe atmosphere, join a local chapter of Toastmasters International.

2. *Talk slowly and clearly.* Before I became a professional speaker, I had one really tough challenge to overcome when presenting. Whenever I got up in front of people, I would talk very fast, like I was rushing to get it done so I could get off stage (and I was!). Once I began training to speak professionally, I started to really focus on this habit. In fact, at one point, it was the only thing I focused on. I didn't care about the content or subject of my speech. I just focused on slowing down my pace. After repeatedly speaking in front of people, everything slowed down for me. Now, when I speak, I naturally speak very slowly and project very clearly. In fact, sometimes, a 60 minute speech feels like 3 hours to me, because the pace seems so slow; yet it's comfortable. This is key to ensure that your audience can hear and digest your message.

3. *Pause periodically.* This is another trick that I learned from some great speakers. By nature, we rush from beginning to end when we do something, without ever pausing for a breath or a break. Pauses at key points throughout your presentations can be extremely powerful. These pauses allow your points to really sink into the minds of your audience and can even add suspense to your talk. Use them often.

"The most precious things in speech are pauses." —Sir Ralph Richardson   [7]

4. *Never, ever read when presenting.* Unless you are reading a passage from a book or a statistic off of a slide, you should try to avoid reading to your audience at all costs. Even if you are new to presenting, prepare as much as possible and try to use the bullet points on your slides as a guide, not a book. Reading is a surefire way to totally lose your audience. You want to connect with them, and when you are reading, your attention is on the literature, not on them. This book's editor Traci Nathans-Kelly has authored a book entitled *Slide Rules: Design, Build, and Archive Presentations in the Engineering and Technical Fields* [8], which provides many strategies for improving your presentation skills for engineers and other technical professionals.

5. *Integrate stories and/or jokes into your presentations.* I know you are an engineer, and many times you are presenting technical information, but that doesn't mean you can't be funny once in a while. Whether it is an inspirational story, a

personal joke, or a real-life example, taking your audience away from the contents of your slides and connecting with them through one of these methods can improve their overall experience tenfold.

6. *Give your audience more value than they know what to do with.* I have always prided myself on giving my seminar attendees tons of value when I speak. I consider "value" to be practical strategies and recommendations that my attendees can put to use immediately in their engineering career and get results—and typically they do. That is how you become a memorable speaker. You change your audience's lives forever by giving them something that they can use regularly to create a better experience for them moving forward. In my opinion, there is no point in getting up in front of an audience if you do not attempt to do this.

7. *End with a bang, but not a loud one.* I like to end my presentations with something inspirational in nature, whether it be a quote, a story, or even a challenge. The end of your presentation is what people are really going to remember. Consider that when you are preparing and delivering it.

Consider this section as an impromptu crash course on public speaking for engineers. Believe it or not, this may be one of the most important courses you take in your career. Remember, the more you speak, the better you get. I have created a complete guide to improving your public speaking skills for engineers, which can be found at www.engineerspeaker.com.

6.9   Confidence Encourages Communication

One of the underlying factors that will determine whether or not you are a good communicator is confidence. You may have observed this with people that you know. They can be very quiet in the workplace, yet when they are in a social setting, they are very loud and outgoing. One of the reasons for this is that they lack confidence in their professional knowledge and experience or in their ability to comfortably communicate it with others.

Confidence is so important to the development of your career because the less confident you are, the fewer risks you are going to take and the fewer opportunities you are going to embrace. You may end up playing it safe during your entire career because you don't think you can do any better. Outrageously successful professionals certainly don't play it safe—they think bold and play big. They sometimes even play bigger than they should and they stumble, but then they get up again and use what they learned to keep moving forward.

> Outrageously successful professionals certainly don't play it safe—they think bold and play big.

If you find that your communication efforts are compromised by a lack of confidence, do everything in your power to give yourself a boost. One way to do this is to think about everything that you have already accomplished in your career. Consider the challenges that you overcame to achieve these things and how good it felt when you did. Think about all of the people out there that may benefit from your ability to communicate better, including your coworkers, staff, clients, and even your family. Join a public speaking group as discussed earlier in this chapter, as this will certainly help you to spark increased confidence.

We are all special in some way, shape, or form. Don't cheat the world from hearing your message, because whether you believe it or not, what you have to say may impact the lives of others. *Be it ONE person or 1000, your message will make a difference.*

6.10 Sometimes Listening Is the Most Powerful Form of Communication

This chapter has focused on communication through speaking; however, listening can be just as powerful, if not more, than mere words when it comes to communicating. Until I attended coaching school and learned how to listen, I was often guilty of selective hearing, which I believe was in large part due to my engineering background.

> "When people talk, listen completely. Most people never listen." —Ernest Hemingway

Engineering professionals are geared toward problem solving; therefore, when we listen, we pay attention for the answers needed to solve problems. From my experience, I have seen too many engineers latch on to these perceived answers, tuning out the rest of the conversation as they mentally solve the problem or look for the next one to tackle.

Another reason people don't listen well is because they like to hear themselves talk. We all do to some degree. We have a lot of thoughts and experiences on our mind, and we want to share them. Sharing our thoughts is great, but engaging and listening to those we are speaking with is important to our relationships, both personally and professionally. Do you find yourself cutting people off before they finish their sentences? This is a fairly common social faux pas that most of us make. We are all anxious to keep moving forward—so much so that we sometimes don't hear important messages that people are trying to tell us, including those from managers, coworkers, clients, friends, and family members.

In coaching school, I learned a very valuable skill called "acknowledging," which is repeating back the words just spoken by another. For example, a client may say to you, "This is our largest project, and it means a lot to us." You would acknowledge the client by saying, "Jim, we understand that this is your largest project and that it means the world to you. That is why we have our best engineers working on the project nonstop." This shows the client that you are listening. And as trivial as acknowledging may sound, it can be extremely powerful in building solid relationships.

> Acknowledging, out loud, the needs of the other person can be extremely powerful in building solid relationships.

How many times have you heard people attribute a problem in the workplace to miscommunication? Do they really mean miscommunication, or do they mean that someone wasn't listening and missed out on hearing what they were supposed to? I believe many times it is the latter. Communication is a two-way street—it has to be. If someone tells you something and you don't listen, what's the point of the whole conversation?

Make it a habit in your career to start listening more. By improving your listening skills, you will set yourself apart from other professionals, and your professional and personal life will be much more rewarding.

6.11 Responsiveness Impacts Reputation

In today's world with all of the readily available technology, people want answers immediately. The success of your career will be directly impacted by how quickly you get them their answers. If real estate is all about location, location, location, then career advancement is all about being responsive, responsive, and more responsive.

Do you remember the last time you e-mailed someone requesting information and it took three days to get a response? How did that make you feel? You were probably very upset, as the lack of responsiveness probably cost you valuable time or money on a project. The next time you have to contact that person, you are certainly going to remember that experience, and maybe down the road, you will just contact someone else in the company who is more reliable or even contact another company altogether.

People just want to be acknowledged. They want to know you are listening to them, and by being responsive to them, you are giving them that recognition. For example, imagine that a client calls and leaves you a voice-mail requesting you to print out a set of design plans that you prepared two years ago because an old project may be coming back to life. You know that it is going to take you at least a day to retrieve the plans, so you have at least two options for responding:

- You can wait until you find the plans the next day and then call the client and request the plans be picked up.
- You can call the client back immediately and acknowledge the message, noting that you received the request and you will return a call with more information in the next day or so once you retrieve the plans.

Even though calling the client back immediately may seem like an extra step, based on my experience with many clients, it is the more favorable option. They will be appreciative that you took the time to call them back, acknowledge them, and show them respect. These are the kinds of things that clients remember when it comes time to hire an engineer for their next project.

> Clients and coworkers are appreciative when you take the time to call them back, acknowledge them, and show them respect. These are the kinds of things that everyone remembers about your work ethic.

Being responsive is important in all aspects of your career, not just with your clients. It's also important within your company, especially if you work for a large firm. You may spend most of your day communicating with coworkers. There will be times when you will be requesting information from other departments, and the rate at which they respond to you will impact the success of your project, and, ultimately, your reputation. Of course, you can't necessarily control how quickly people respond to you, yet I find that if you are responsive yourself, people tend to return the favor. One approach you can take is that when you request information from someone, attach a time deadline to it, but do so in a nice way.

> If you don't respond in a timely manner, next time they might not be calling you.

Please, take this advice to heart. I have had people call me and say that they planned to hire my firm because someone else told them how responsive we are. You would think that if someone calls or e-mails you, it would be important to get back to them in a timely manner; however, many professionals don't do this. It often comes down to being organized in dealing with your e-mail and phone messages (which I will discuss in Chapter 8).

Your professional responsiveness can greatly affect your reputation as well as that of your company. I know you are busy, but taking the time to respond to people is critical to your success. If you don't respond in a timely manner, next time they might not be calling you.

6.12 Key Points to Remember

1. The key to good communication on project teams is to make sure that everyone is on the same page, whether team members decide to use the phone, e-mail, or any other form of communication. Be sure to establish communication guidelines with your team as early as possible to avoid extraneous exchanges and conflict.

2. Whether you are preparing a design for a new project or overseeing the construction, it is important to communicate with the clients as well as the end users of the project as early on as possible. Addressing their needs in the design may yield much more practical, useful, and efficient results.

3. To successfully relay technical information to others, think about the information you are presenting from the other person's perspective. Use analogies to put your information into terms that your audience will understand.

4. Regardless of the pressures from your clients or the stress imposed by deadlines, always be honest in your career. Telling people what they want to hear may be easier in the moment, but it will catch up with you and negatively impact your reputation.

5. Remember, the way that you say something is just as important as what you say. One way to be respectful of people is to ask questions of them rather than always telling them what to do. Asking questions of people will get their "buy-in" or commitment to a certain outcome.

6. Public speaking can help you advance your career in so many ways including increased confidence, improved communication skills, new business development opportunities, and facilitation of project approvals. Join a local Toastmasters group today to start to improve your public speaking skills.

7. Consider taking some formal training to improve your presentation skills, possibly joining a chapter of Toastmasters International or a similar organization.

8. Listening is an important part of communicating. To improve your listening skills, try waiting until people are finished talking before you start, and also acknowledge what others say to you by rephrasing it and repeating it back to them.

9. Responsiveness in your career will greatly affect both your reputation and that of your company. Take the time to respond to people. If you don't have an answer right away, at least let them know that you have received their request and you are working on it.

A Boost from Your Professional Partner

Are you scared to death of public speaking? Join the club—most professionals are terrified of speaking in public. Unfortunately, for most people, the ability to speak in public and clearly present their ideas can be a career game changer.
Remember that in order to create an extraordinary career, you'll have to do the opposite of what most people do.
You can do it!

Your Professional Partner,
—Anthony

References

[1] M. Caroselli, Leadership Skills for Managers. New York: McGraw Hill, 2000, p. 71.

[2] D. Goleman, A. McKee, and R. E. Boyatzis, Primal Leadership: Realizing the Power of Emotional Intelligence. Boston, MA: Harvard Business Review Press, 2002.

[3] Z. Ziglar and LLC Meadow's Edge Group, Life Wisdom: Quotes from Zig Ziglar: Inspire to Be Great! Nashville, TN: B&H Publishing Group, 2014.

[4] D. Olive, The Quotable Tycoon: An Irreverent Collection of Brutally Honest and Inspirational Business Wisdom. Naperville, IL: Sourcebooks, 2002, p. 203.

[5] E. O'Brien, How to Make Powerful Speeches: A Step By Step Guide to Inspiring and Memorable Speeches. Dublin: The Reluctant Speakers Club, 2014.

[6] Statistic Brain, Fear of Public Speaking Statistics. Available at http://www.statisticbrain.com/fear-of-public-speaking-statistics/ (accessed on February 5, 2014).

[7] M. Cohen, Penguin Thematic Dictionary of Quotation. New York: Penguin Group, 1998.

[8] T. Nathans-Kelly and C. Nicometo, Slide Rules: Design, Build, and Archive Presentations in the Engineering and Technical Fields. Piscataway, NJ: Wiley-IEEE Press, 2014.

7

The Ability to Network

The successful networkers I know, the ones receiving tons of referrals and feeling truly happy about themselves, put the other person's needs ahead of their own.
—Bob Burg [1]

7.1   What Is Networking and Why Is It Important?

Networking is a word that you will hear often in your career as an engineering professional. Everyone has a different definition of what networking means. I like to keep it simple and define networking as two words—building relationships. Bob Burg quoted at the beginning of this chapter is the best-selling author of *The Go-Giver*, a book that stresses the importance of a mind-set of "giving" when networking instead of "getting" [1].

If you want to have an extraordinary career, it is imperative to build solid relationships, not only in your industry but also in every aspect of your career and life as a whole. While building relationships is one of the most important elements to career success, it is something that many engineering professionals struggle with for a multitude of reasons. Whether it's our technical nature, lack of confidence, being too busy, struggles with language, or just a lack of desire to converse with others,

Engineer Your Own Success: 7 Key Elements to Creating an Extraordinary Engineering Career,
First Edition. Anthony Fasano.
© 2015 The Institute of Electrical and Electronics Engineers, Inc. Published 2015 by John Wiley & Sons, Inc.

there are many obstacles to mastering this critical skill. Most engineering programs that I have looked at don't offer a class on building relationships, which I intend to work to change that because I believe engineers are at a huge disadvantage without this skill.

The reason I define *networking* as "building relationships" is because that is exactly what you should be doing when you network. Many people think that networking is going to a social outing, collecting business cards, and trying to secure clients on the spot. That's not what it is. In fact, networking or building relationships is truly defined by what you do after that first meeting.

> Networking is truly defined by what you do after your initial meeting with someone.

I recommend that your goal in networking be to build strong relationships, because successful careers and enjoyable lives are built on solid relationships. We are constantly creating associations with our coworkers, clients, team members, supervisors, family members, and friends. That's what we do on a regular basis, so it must be important. It's also a key ingredient to our success because the more relationships we have, the bigger our network will be, and, in turn, more opportunities will be available to us and to our company. It's not just the size of our network that counts, though; the quality of our relationships will also have a big impact on our success.

> "The quality of your life is in direct proportion to the quality of your relationships."
> —Anthony Robbins   [2]

In this chapter, you will learn how to overcome personal obstacles to networking and build strong, lasting relationships. Whether you are looking to bring business into your firm, find a new employer, or create more enjoyment in your career, the ability to network effectively can help you.

7.2   Secrets to Building Lasting Relationships

To help you start out, I have some great tips on how to build strong relationships, including some secrets that I have learned through my experience. I don't refer to them as "secrets" to imply that I want to hide them from people; I call them that because you would think they were secrets since so many people don't know about these strategies or else just fail to use them.

The best way to build relationships is to focus on building personal associations or friendships. Whether it is a business relationship or not, take some time to get to genuinely know people. Learn about their families and their interests outside of work. Show interest in them as a people.

> The following networking techniques, so hard won in some Western cultures, are the very foundation for business in other cultures, such as Japan:
>
> - Ask questions related to their past like "Did you grow up around here?"
> - Talk about your family and then ask a question like "Do you have children?"
> - Ask exploratory career questions like "What made you want to become an engineer?"

7.2.1 Their Interests Should Interest You

From my experience in the United States, people approach networking like they are on a mission; they want to talk to the person simply to get what they want and then get away from them. Not only is this a terrible way to try to develop your career, but it also is a terrible way to live. Networking is about building a connection with the other person beyond just finishing a project or making a sale. I am not recommending that you spend hours talking with people because that is not realistic; but I will say that spending an extra 5 minutes speaking with someone about that person's interests can greatly impact your long-term relationship.

> Networking is about building a connection with the other person beyond just finishing a project or making a sale.

Be sure that as you take the time to get to know people that you remember what you learn about them. If they tell you information about their hobbies or mention their children's names, make a note of it on the back of their business card or in a notepad immediately after the conversation and eventually transfer this knowledge into your contact database. While remembering people's names is an important aspect of relationship building, if you can also remember their children's names or their favorite vacation spot, you will garner an amazing amount of trust and respect from that person and quickly develop a very strong relationship. There are several books on networking listed in Appendix A of this book including Cialdini's *Influence* [3] that provide pages of powerful details on networking strategies.

7.2.2 Listen to Others

Another tool that you can use while building relationships is the powerful one of listening, which was discussed in Chapter 6, Section 6.10. We often get the urge to cut people off when they are talking because we want to contribute to the conversation. Try waiting, as hard as it may be, and let the other person finish before you jump into the dialogue. When I started to do this, the quality of my conversations rose dramatically. I felt like I was connecting more than when I used to cut off that person's thoughts. I have found that the art of waiting is an extremely difficult thing for engineering professionals to do, and I think it may be partially attributed to our inherent nature to want to quickly solve problems and produce answers. That is why we are always trying to jump in and give an answer or solution long before a question is even asked.

Let the other person finish. If you want to be really extreme, try to leave one or two seconds of silence after the person finishes talking and before you start responding. While this may be difficult at first, once you master the technique, you will get much more out of your conversations and, ultimately, your relationships.

Your body language, including your facial expression, is also very important when building a relationship with someone. If you always seem overly reserved or if you are not accustomed to showing much emotion, people might not enjoy talking with you and ultimately try to avoid you. There are certain people that I just won't converse with because of their negative attitude and expressions. Along the same lines, people will be more attentive when you are speaking if you stand or sit up tall and smile when talking to them. It really does make a difference.

7.2.3   Relationship Value Is a Two-Way Street

Another great way to build relationships with people is to give them value in whatever ways you can. Always be a resource for your contacts. For example, if you are developing a relationship with an architect who designs schools specifically with an emphasis on green design and sustainability, send that personal relevant article on that topic when you happen to run across it. Whenever I read a good article or book, I try to think of people I know who would benefit from the same and send them a note directing their attention to it. Even though you may not have written the article or book, they will associate the value that they receive from it with you because you gave it to them. Also, once in a while, ask a contact out to lunch, and during the meeting, ask if you can be of any help career-wise or with a project.

Again, be a resource to people and put their needs before yours. Remember what I said about networking being about what you do after your first meeting? These are simple things that so many people don't take the time to do when building relationships.

So, if you were using the excuse that you didn't know how to network, you can't use it anymore. Start using some of the strategies provided in this section to begin building strong, lasting client (and personal) relationships. Not only do these guidelines yield better career opportunities, but by really getting to know people, your career will truly be more enjoyable. A big part of having an extraordinary career is being able to enjoy it, and the tips in this chapter will help you to do just that.

7.3   Network in Your Industry Through Professional Societies and Organizations

Taking your newly acquired networking skills into practice is the next important move. It's essential to network within your industry for many reasons. Networking becomes key to your success at many points in your professional life: you may need to partner up on a project with other engineering firms or professionals; you may need assistance from your network if you lose your current job; you may want to jump-start a stagnant career by contacting mentors or possible new project modes.

The best place to network with other engineering professionals is through professional engineering (PE) societies, groups, and organizations. I actually founded a private career development community for engineers known as the Institute for Engineering Career Development (www.engineeringcareerdevelopment.com) to foster networking and support between engineers, because I feel so strongly about building solid relationships.

Likewise, all engineering disciplines have specific professional societies. As a civil engineer, I am a member of both the American Society of Civil Engineers and the National Society of Professional Engineers. Professional societies and organizations are usually nonprofit organizations with the goal of advancing visibility of a particular profession, through increasing the public's interest and attracting members. These organizations typically hold regular meetings that will allow you to get to know other professionals in your area. The meetings usually provide an educational program along with a networking component, which makes them very valuable to your career advancement.

Top 10 Engineering Associations by Membership

Association	Website	Membership
American Institute of Aeronautics and Astronautics (AIAA)	http://www.aiaa.org/	35,000 members
American Institute of Chemical Engineers (AIChE)	http://www.aiche.org/	45,000 members
American Institute of Mining, Metallurgical, and Petroleum Engineers (AIME)	http://www.aimehq.org/	150,000 members
American Society of Civil Engineers (ASCE)	http://www.asce.org/	145,000 members
American Society of Mechanical Engineers (ASME)	https://www.asme.org/	130,000 members
Institute of Electrical and Electronics Engineers (IEEE)	http://www.ieee.org/index.html	425,000 members
Institute of Industrial Engineers (IIE)	http://www.iienet2.org/	14,000 members
National Society of Professional Engineers (NSPE)	http://www.nspe.org/index.html	35,000 members
Society of American Military Engineers (SAME)	http://www.same.org/	27,000 members
Society of Women Engineers (SWE)	http://societyofwomenengineers.swe.org/	20,000 members

Not only do I recommend joining your industry's society, I urge you to get involved as a volunteer by serving on one of its boards or committees. This will help you to build your leadership skills as well as establish an excellent reputation in your industry. Most companies encourage their employees to join and participate in these organizations, and some may even reimburse employees for the annual membership fees and fees associated with events.

The reason I encourage you to join and participate is because many engineering professionals enroll but then fail to actively involve themselves with the organization. They are members, but they never go to any meetings. They may claim to be too busy working; however, what they fail to see is that attending these events may be just as important to their career as the actual technical work.

Don't just join professional associations; volunteer for a leadership position or committee chair. The skills you will develop and the confidence you build will be invaluable to your career and personal development.

> Don't just join professional associations; volunteer for a leadership position or committee chair. The skills you will develop and the confidence you build will be invaluable to your career and personal development.

It is so easy to get wrapped up in projects, paperwork, and the day-to-day routine of work. Yet if you want to have an extraordinary career, you are going to have to broaden your horizons. You will need to step out of your technical shell and connect with people by attending these events and even giving presentations at these meetings. You may even get the opportunity to travel to a conference and see some new and exciting places. Some of these conferences are held in great travel destinations all over the world like Maui, Miami, Chicago, New York, Panama City, Anchorage, Brussels, Amsterdam, Beijing, Cape Town, and Madrid—all the more reason to attend. Professional organizations provide you with both technical resources and a forum for building a strong network in your industry. It's up to you to take advantage of what these organizations have to offer. As the old saying goes, you get back tenfold of what you put in, so get busy networking.

7.4   Finding and Developing Project Leads Gets You Noticed

If you are a consulting engineer, make no mistake about it: bringing new work into your firm is the fastest way to advance your engineering career. Whether or not you have a PE license or a master's degree, if you bring in contracts, people will notice. For information on how to obtain your PE license or a similar certification in your country, see Chapter 4.

Every top engineering executive that I have spoken with has told me that if you want to get ahead in this industry, bring work into your firm. New projects are the lifeblood of any company; without them, a company won't survive. So, while all of the elements in this book are important, building strong working relationships can influence your ability to contribute to the organization's bottom line.

> Every top engineering executive that I have spoken with has told me that if you want to get ahead in this industry, bring work into your firm. New projects are the lifeblood of any company; without them, a company won't survive.

The two most important things to understand when attempting to develop new business opportunities is *who* your clients are and *where* you can find them. For example, as a civil engineer in the land development arena, my clients were property developers, real estate agents, and also government entities that were planning to make improvements to properties and facilities. Once you are clear on this, then you can determine where these people spend their time; this allows you to plan ways to connect and build relationships with them.

Make no mistake about it: bringing new work into your firm is the fastest way to advance your engineering career.

The strength of this two-step approach lies in reaching out. Many engineers aren't exactly clear on who their clients are, or they don't spend their time in the right places to be able to network with their prospective clients. For example, some engineers spend all of their time networking with other engineers through professional associations; however, engineers are not their own clients, usually. While you should attempt to be active with your colleagues in organizations like this, you must use your time wisely and also focus on the avenues that will help you to achieve your goals; and for those of you reading this that want to become partners in an engineering company or even own your own business, business development is critical.

Assuming you are clear on what demographic you serve, the next step is getting in front of them. Again, in an effort to be efficient and smart with your time, it is best to explore avenues where you can get in front of multiple prospective clients at one time. I will use the same example as earlier in this section, a civil engineer who works in land development with real estate developers and agents and municipalities as the main prospective clients. Here is a list of possible avenues where an engineer in this situation could gain access to prospective clients:

1. Join online networking groups (i.e., LinkedIn groups) that are populated mostly by your prospective clients. Then, get engaged in those groups and provide consistent valuable information that will be helpful to the group members. Don't try to sell yourself; simply be a resource to them.

2. Attend conferences where prospective clients would gather and maybe even invest in a booth (with your company's approval, of course). Be sure to hand out materials at the booth that provide information to your prospective clients, not just sales brochures. For example, maybe you or a fellow employee writes an article, "Top 5 Things to Consider When Purchasing a Property to Develop." Add your company's name, website, and contact information on each page of the article. If you are representing your firm, then you need to get full approvals and consent for this type of activity.

3. Advertise in or submit a technical paper to magazines or on websites where your prospective clients will read or visit. Don't focus solely on engineering journals; advertise in and submit articles to publications that your prospective clients will be reading.

4. Consider joining an association or group that serves your target market. For example, maybe there is a green building group or committee in your community that is open to everyone. By becoming active in the group, not only will you get to network with prospective clients, but also you will become more knowledgeable on their industry and elevate your expert status.

The key to developing strong relationships and eventually project leads is following these steps listed above to get you in front of the right people and then implementing the relationship-building strategies we discussed earlier in this chapter. From my experience coaching engineers, if you do these things, you will be light years ahead of most engineers in today's corporate world.

7.5 Opportunities Have No Limits

While networking within your industry may be a great place to start, you will eventually have to move beyond these limits into the world at large—speaking with nonengineering professionals, such as local citizens, contractors, government officials, as well as existing and prospective clients. It is important to build relationships with your clientele, but there are other community members who are end users of our projects that are too often overlooked.

Just like there are industry-specific professional societies, there are also other general networking groups that will allow you to develop relationships within the community and with professionals in other industries. Developing relationships with engineers and clients is great, but limiting yourself to these sectors may cause you to miss other exciting opportunities. Some groups that may be beneficial to join outside of your industry include business referral groups, Rotary, Toastmasters, and local chamber of commerce groups. There are also some engineers who volunteer in local government and later realize that this is a great way to learn about new projects like schools, housing developments, or roadway improvements. They can then take this information back to their firms.

Joining groups like these can benefit you and your company in many ways. By networking with community members, you may become more in tune with what the community needs, which could eventually improve the quality of the design and construction of your projects. The more well known you become in your community, the greater the likelihood that you will be the one they call when a need arises. I know I said this earlier, but again, as you get to know people in the community where you work, your career becomes more enjoyable because you are not just working to improve the lives of people, you are working to improve the lives of people you know.

> As you get to know people in the community where you work, your career becomes more enjoyable because you are not just working to improve the lives of people, you are working to improve the lives of people you know.

I have always wanted to advance my career as quickly as I could, so early on, I joined a business referral group. This is a group where they permit no more than one person in each profession into the general membership. The goal of the group is to exchange business referrals at their weekly breakfast meetings. I joined the group because my brother had recommended it, but I wasn't quite sure how it was going to benefit me because the group members included a financial advisor, an accountant, a payroll service company, and a house cleaning company, among others. I was only thinking in terms of people that might need my services and not necessarily thinking of the bigger picture as to how these relationships could help my career and my employer. One evening, some fellow coworkers and I participated in a public hearing for a high-profile streetscape project that we were working on. The citizens of the town had formed a civic association and wanted to share thoughts on the project at the meeting. We had already had a few meetings with this association in the past, and they were very hostile toward our design team.

> Be bold, be brave, be yourself, and the opportunities will just keep coming.

As soon as I walked into the room that evening, I saw the accountant from my business referral group. I went up to him and we exchanged a warm greeting. I found out that he had taken over as the president of the civic association. Needless to say, the evening went much smoother than we had thought it would, largely due to my existing relationship with the accountant. You may say, "Wow, what a coincidence." However, I would argue with you and say it wasn't a coincidence at all. By joining the referral group, I opened myself up to more opportunities and met people from different business arenas. I put myself in a position to succeed, and succeed I did that night. Not only did the meeting go extremely well, but also my supervisor noticed that it was my relationship with the accountant that helped the project move along. Our company also looked very good in front of our client, the municipality.

Don't be too quick to judge or shy away from joining a group just because you feel that the members are not all engineers. Be bold, be brave, be yourself, and the opportunities will just keep coming.

7.6 You Are Never Too Young (or Old!) to Network

Many professionals have the idea that only the more experienced professionals have to network and build relationships. Typically, early on in your engineering career, you will be asked to stay in front of the computer and crank out calculations, create designs, or perform site inspections. Waiting until your later years to network is a great approach to take if you want to have an average, ordinary engineering career. Fortunately for you, this book is about having an extraordinary career, and in order to do so, you need to let go of that belief that you are too young to network *right now*.

Regardless of what you are asked to do, you must take it upon yourself to incorporate networking into your schedule. You are never too young (or old for that matter) to do anything to advance your career, especially networking. In fact, since so many people believe that fallacy, if you do start networking while you are younger in your career, you will set yourself apart from other professionals (see Section 4.2) who subscribe to this outdated theory, and you will put yourself in a position to succeed.

That's what advancing your career is all about—doing things that make people say, "Wow. That person is a real mover and shaker." Always remember that networking is about building relationships, which is what you have been doing your whole life. The difference now is that you are doing this in a professional setting.

If you are a younger engineer, when you do some of the things that I have recommended such as joining networking groups and going out to lunch with your new connections, you may hear some backlash from other people. They may say you are "too young" to start worrying about networking or that you should be focusing on doing your job. Professionals that listen to these kinds of statements go on to have mediocre careers. Those that don't listen to these mudslingers rise to the top quicker than most, in my experience. The fact that you are reading this book tells me that you are determined to have an exciting and successful career. Start building relationships as early on in your career as possible, regardless of what anyone tells you.

> Start building relationships as early on in your career as possible, regardless of what anyone tells you.

7.7 Overcoming Low Confidence and Language Barriers

Based on my experience coaching engineers, the two biggest obstacles that hold engineers back from networking effectively are lack of confidence and language barriers. If you practice engineering in a language other than your primary one, it is totally understandable for you to lack confidence when networking. The best way to overcome this is to start taking small steps to build your confidence. For example, rather than trying to start a relationship with someone you don't know, begin by further developing existing relationships. Maybe it's an architect that you had teamed up with on a previous project or an existing client. Call that contact and ask for a lunch meeting to catch up. This will give you a good opportunity to make a strong connection with that person.

By initiating some relationship-building activities with people you already know, you will really boost your level of self-confidence. If you do reach out to an existing contact and have a successful conversation or meeting, make a note for yourself of any specific thing you may have done during your meeting that set the tone for a great interaction. Then, when it's time to start networking with cold contacts or people you don't know, you will have two items to build on: prior experience and some proven methods that you can utilize. From a confidence standpoint, knowing that a certain strategy has worked well for you in the past will make it much easier for you to use it a second time.

Effectively Learning From Past Experience

Whenever you do something successfully, it makes it easier to do it again. In the case of networking, you can build on previous successful conversations. This is primarily related to the confidence that you gain from these conversations. The next time you have to converse with someone you don't know, you draw on this confidence by visualizing the previous success. This visualization will bring that confidence to the forefront. It will also allow you to remember the kinds of questions and terms that you used in that situation so that you can repeat them again.

Language barriers also present a huge obstacle for engineers. Many engineers that conduct their business in a language that isn't their primary mother tongue are very fearful to network with others, because they feel their command of the second language will affect the quality of their conversations. This is definitely a challenge because their ability to speak and understand English directly affects their confidence. Along the same lines, many professionals get lazy and use slang words in the workplace. This can reflect very poorly on your reputation as doing so may prevent people from taking you seriously or viewing you as immature.

The best advice I can give anyone wishing to overcome any language barrier is to join a local Toastmasters group as I discussed in detail in Section 6.7. Many professionals join Toastmasters to further develop their speaking and language proficiency, build their confidence, and develop their public speaking skills in the process. Toastmasters forces you to exercise your language arts and speaking skills in a professional setting.

7.8   How to Deal with a Boss or Supervisor Who Is Holding You Back

Throughout my years coaching engineers, many engineers of all different experience levels have asked me the same question: "What do I do if my boss or supervisor is stunting my engineering career development?" Most of the engineers who ask me this question are facing the same challenge; their boss will not let them do tasks or take on responsibilities that they believe they are ready for. Some of the tasks or responsibilities many of these engineers mention include client correspondence, presenting at public or client meetings, having leeway to make some decisions without their boss's approval, and increased responsibility with respect to managing staff.

This is not a difficult topic at all for me to write about, because I faced this exact challenge for a good part of my engineering career. In fact, as ironic as it may be, this challenge in my career is the reason I ultimately started the Institute for Engineering Career Development and wrote this book.

I was a very motivated and driven design engineer. I worked at a very rapid pace all day, every day. However, once I reached a certain experience level, in my case about five years into my career, I hit a wall. My supervisor at the time didn't think I was ready to take the next step (actually he wasn't ready to give any of his work away), and therefore, I was stuck performing tasks that were well below my skill level. I became totally disengaged in my engineering career. At times, I felt that my boss was threatened by my rapid advancement, and he didn't want me to advance anymore. This was the most frustrating time in my career—by far.

I considered finding a new job. I was so disengaged, though, that I decided first to explore other career paths. I had always been very interested in helping other engineers advance their careers, possibly through career coaching or something along those lines—so I looked into it. I ended up enrolling in one of the top executive coaching schools in the world—the Institute for Professional Excellence in Coaching (iPEC). I remember sitting on the couch with my wife, Jill, telling her, "I think I want to be a coach." Within a week's time, I was signed up for iPEC and clearly remember driving four hours to the first weekend training session. That car ride was some of the best hours of my life. There was a sense of freedom running through my body, as I knew I had found my calling.

The iPEC program was a 500 hour intensive night program, so I was still able to continue in my "boring" engineering job at the time. Once I realized I had a real talent (and now the skills) for engineering career coaching, I started my company Powerful Purpose Associates and have been helping engineers ever since. I am not telling you to enroll in coaching school if you are facing this challenge, but from my own experience and through helping other engineers with this challenge, I would give the following advice:

- *Be sure to ask your boss for more responsibilities every chance you get.* I am not saying to camp out outside of your boss's office asking for more responsibilities all day every day (I tried that and it didn't work), but I am telling you to ask at least a few times a week in different ways. If there is a staff meeting to discuss project workload, make it clear that you are open to helping out with tasks above and beyond the ones currently on your plate. If your boss tells you that he or she is swamped, volunteer to take some of the workload. If the boss tells you that you won't know how to do it, say that you are a quick learner. Be aware, too, that you should document your attempts. Confirm in e-mails your offers to help, which can become important during annual reviews that examine your work initiatives.

- *Ask your boss if he or she would teach you some new roles and responsibilities.* Similar to the one above, but instead of just asking for more responsibilities, ask the boss to teach you, in a mentoring relationship. Sometimes, managers feel threatened to give people some of their work or billable time, yet when it comes to helping or mentoring, they are more than willing to give of their time.

- *Tell your boss that you do not feel challenged enough at work.* When doing this, say something like "I would like to take on some new challenges. Would you mind if I asked some of the other managers if they have work that I can help them with?" And I add this warning: this approach can end up in one of two ways. One, rather than losing you to another manager, they will start giving you more responsibilities (slowly—but it will be better than nothing). The other option is that your idea is given approval; you will get to work with another manager and broaden your horizons. However, be prepared to do that extra work.

- *Don't be too quick to leave the company you are working for.* If the company you are working for is a good company, exhaust all of your options there before leaving to go somewhere else. Talk to your human resources (HR) department about the possibility of transferring to another department or nearby office or being reassigned to a new supervisor. When you bring this to the attention of the HR department, those in that office may realize this isn't a one-time occurrence with your boss, which may bring about some changes in the company structure. It may be both in your and your employer's best interest to address this problem and create a better work environment overall. It can't hurt to ask before leaving.

- *Use your downtime to improve your weaknesses.* If you are facing a challenge similar to mine, you may have free time on your hands. While you might want to take on more work responsibilities, if that's not an option, use the time to improve some of your weaknesses or build new skills. Maybe you always wanted to be a better public speaker. If so, and if your boss doesn't allow you to present at workplace meetings, join a local speaking group in the evenings—that's what I did.

- *Look for the opportunities in the situation.* Something I always tell members of my Institute for Engineering Career Development is to look for the opportunity in every situation in your engineering career. If you look, you will always find one, just like I did. In fact, one of the Institute for Engineering Career Development members was working in a firm in a field outside of the field he wanted to practice it. He ended up getting laid off and saw it as opportunity rather than disaster. He found the job he always wanted soon after. You can read his story, called "Realizing a Dream of Becoming a Structural Engineer" in Chapter 12.

- *Recognize when it's time to move on.* If you have exhausted all options in your current situation and nothing gets better over time, then look for a job with a different employer. This is a last resort, but a very important one. If you stay in a bad situation too long, your engineering career development could seriously suffer, and your career could end up being stagnant for years.

7.9   Interoffice Politics and Workplace Relationships

In addition to employers expecting engineers to work too many hours, one of the biggest negative aspects of the corporate world is the negative energy that can develop in an office due to the competitive nature of the corporate world. Two engineers who have very similar personalities, who might make for great friends and colleagues, are often turned against one and other in an effort to see who can earn a higher salary or a loftier title.

Please don't let this negative approach take over your career development efforts. It is this attitude that incites interoffice politics and bad relationships. While it's important to stand out from the crowd, I believe that by developing the skills discussed in this book, you can take a very holistic approach to career development. You can advance as far and as fast as you would like, regardless of what anyone else in the industry is doing. You don't have to compete with others. Set clear goals and develop and implement the skills that you need to achieve them in your own career.

> You can advance as far and as fast as you would like, regardless of what anyone else in the industry is doing.

When you start to worry about what the next person is doing and how you can do it better, that is when you will falter. Such a mindset forces you to abandon your own strengths and try to be someone that you are not. If someone in the office is constantly trying to beat you out, ignore it and focus on achieving your goals. Let negative energy roll off your back, and certainly don't return any of your own; that approach only creates a never-ending circle of negative energy.

In fact, to the contrary, I would recommend helping your colleagues to advance as much as possible. If they are under the gun for a deadline, offer to stay and help them finish the project. If they are studying for a licensing exam, offer to take some of their workload from them. I found that people who take this approach are often happier in their engineering career and still achieve their goals in the process. Don't let the politics of interoffice relationships weigh you down, because they will—if you let them. It's up to you to maintain positive relationships that will support your goals and your happiness.

> "Thousands of candles can be lit from a single candle, and the life of the candle will not be shortened. Happiness never decreases by being shared."—attributed to Buddha   [4]

7.10   Monitoring and Controlling Your Professional Image in Social Networking

We discussed in Chapter 6 all of the different ways to communicate in today's world. Those lessons learned also apply to networking opportunities. In addition to all of the ways of building relationships discussed earlier in this chapter, there are also now many places for networking online known as "social networking." Social networking is a form of networking where people share information and build relationships through online communities. At the time this book was published, the major social networks included Facebook®, LinkedIn®, Twitter®, Google Plus®, and YouTube®. There are also more specialized private communities like my Institute for Engineering Career Development. I definitely recommend that you engage in some form of social networking, as it is an extremely powerful tool in building relationships and, ultimately, in advancing your career.

7.10.1   Controlling Your Facebook, Twitter, and Google+ Messaging

Most of you reading this book probably already have a Facebook account. While Facebook is not extremely helpful with respect to engineering career development, it can be harmful to you. I recommend keeping your Facebook profile clean and adjusting your privacy settings so that those not connected to you cannot view your profile. Untag yourself as needed in photos or messages that others may have of you. Many companies when interviewing prospective employees will review their Facebook profiles as part of their candidate review process. You do not want a vulgar post by you or even one of your friends on your page to exclude you from landing a great job.

As time goes on, Facebook, Twitter, and Google Plus may become more helpful in career development. Right now, they are mostly for personal use, although professionals and organizations are starting to use these social media sites for business more and more. Most engineering organizations start Facebook pages to keep their members up to date on events and other information.

Along with the benefits of social networking, there are also several potentially harmful aspects of being involved with this type of interaction. The first and foremost drawback is the amount of time that you can spend on these websites. Social networking is powerful, but it can also become addictive, especially sites like Facebook where you might log in just to check something and end up staying on it for hours. Many employers block Facebook from employees during work hours to avoid this problem, but it can still suck you in at home and take away from your personal time or exam study time, which can impede your career advancement.

Another aspect of these sites that you must be aware of is that, depending on your privacy settings, *your posts may be visible to everyone that looks at your profile.* You can set some privacy parameters in place, but always be mindful that what you type may become public knowledge. In fact, if you "Google" your name, most likely your LinkedIn profile will come up if you have one. I once knew an engineering professional who had posted on his LinkedIn profile that he was pursuing a new job in the engineering industry. Meanwhile, he didn't realize that his current employer could see his profile. Ouch.

Also, please keep all of your social media profiles clean. Think of them as your professional online résumés, because basically that is what they are. As far as Facebook or other similar sites go, I am not going to tell you what to put on your profile. I will simply tell you to be aware of whom you are connected to and consider the consequences of each and every post you make. Untag yourself in photos, if necessary. Many of the same strategies are used whether you are trying to find a job or you are simply building your network.

7.10.2 Maximizing LinkedIn

I recommend that you utilize LinkedIn as you primary social media outlet when it comes to your professional development. LinkedIn is a social networking site that is totally dedicated to business and is used primarily for professional networking. It was founded in 2002, and by 2013, it had over 259 million users from around the world [5]. Executives from most of the Fortune 500 companies use LinkedIn. When you hand people your business card today, most of them will look you up on LinkedIn, so be prepared.

Many professionals fail to realize how powerful a website like LinkedIn can be in terms of advancing your career. It can help you find another consultant to partner with on a project, meet new clients, find employment, and so much more. LinkedIn has a feature called Groups where you can join any one of thousands of industry-specific groups. Once a group member, you can ask questions of other professionals or engage in the many discussions on different industry topics. I have been able to gain valuable career development information for myself using the discussion segments of the Groups. For example, I have received career advice, found guest speakers for The Engineering Career Coach Podcast, and connected with prospective clients. You can also start a group yourself like I did (The Engineering Career Coach: Engineering Advice and Discussions; the group is a private group—for engineers only because my last two groups became overrun with recruiters). You can even start a private group for your organization or society, if you desire.

You must set time limits for using these sites, especially LinkedIn, which is probably available for your use at work because of the nature of the site. Set aside 30 minutes a day or so where you can go into LinkedIn, connect with some people, and get involved with a few discussions. It may take more than 30 minutes for the first few weeks to get your account and network set up, but be sure to get back to a 30 minute or less schedule after the initial preparation of your account is completed. See Chapter 1 of this book where I give seven steps for setting up your LinkedIn profile.

Whether you are a young engineering professional or an experienced one, *social networking sites can provide unlimited opportunities if you use them wisely.* I recommend you do use them and seek training beforehand to be sure that you are comfortable utilizing them. These sites can make it easy to do things like obtaining specific information about someone and meeting people around the world. Such ease of opportunity was nearly impossible to do just a few short years ago.

Please use the information in this chapter to start building solid, influential, and rewarding relationships today.

7.11 Key Points to Remember

1. Don't think of networking as collecting business cards; think of it as building relationships.
2. The best way to build relationships is to focus on building personal associations or friendships. Take the time to get to know people and remember what you have learned about them, including their personal interests and hobbies.
3. The best place to network with other engineering professionals is through professional engineering societies and organizations. Join a professional society, attend meetings, and get involved as a volunteer by serving on their boards or committees.
4. Push yourself outside of your comfort zone and join general networking groups. This will help you develop a well-rounded network and allow you to become much more involved in the local community.
5. When attempting to develop new business, be sure that you are clear of who your clients are and where to network effectively.
6. Don't assume that you are too young in your career to start networking. Start building relationships as early as possible.
7. Start networking with people that you already know until you are confident enough to approach new acquaintances.
8. To overcome language barriers, join a Toastmasters group to build confidence and develop public speaking skills at the same time. Improving these skills will make you more comfortable when networking.
9. If your boss is hindering or unaware of your career development efforts, address the situation as quickly as possible to avoid years of stagnation.

10. Focus on developing your skills, achieving your goals, and helping others to do the same. Do not subscribe to the typical corporate mentality that you must put others down in order to advance.

11. Utilize social networking sites, specifically LinkedIn, to build relationships with other industry professionals. Use the LinkedIn Groups feature to ask other professionals for guidance or advice on career decisions. Be aware that your posts may be public and accessible to everyone, depending on your security or privacy settings.

12. Use the information you have learned in this chapter to start building relationships and bringing in new business to your company. This is the fastest way to get noticed in your firm and advance your career.

A Boost from Your Professional Partner

Here is a brief but powerful career advancement tip.
Three words: GET OUT THERE!
You will be amazed by the career opportunities that will fall into your lap by just getting out there and networking. I can promise you that nothing exciting will happen if you don't!
I'm just giving you a gentle PUSH.

Your Professional Partner,
—Anthony

References

[1] B. Burg, The Go-Giver. New York: Portfolio, 2007.

[2] A. Robbins, Awaken the Giant Within. New York: Simon & Schuster, 1992.

[3] R. B. Cialdini, Influence: The Psychology of Persuasion. Needham, MA: HarperCollins, 2007.

[4] Tiny Buddha [Online]. Available at http://tinybuddha.com/wisdom-quotes/thousands-of-candles-can-be-lit-from-a-single-candle-and-the-life-of-the-candle-will-not-be-shortened-happiness-never-decreases-by-being-shared/ (accessed on April 14, 2014).

[5] LinkedIn Corporation, "About LinkedIn." Available at http://press.linkedin.com/about (accessed on January 21, 2014).

8

Stay Focused, Organized, Productive, and Stress-Free

Efficiency is doing things right;
effectiveness is doing the right things.

—Peter Drucker [1]

8.1   The Three Rules to Time Management and Work–Family Balance

Through my Institute for Engineering Career Development, I have had the opportunity to provide career coaching and guidance to hundreds of engineers. If there was one challenge that I have seen surface repeatedly for engineers, probably 75% of those that I coach, it is time management and, ultimately, work–family balance. Some people refer to it as work–life balance, but most engineers that I work with say to me, "I want to be a successful engineer, but I would also like to be able to spend more time with my family away from the office."

This is a very common and interesting, almost ironic, problem. Most engineers work hard to earn a high enough salary to be able to support their family; however, to do so, they can only spend very little time with their family. What's the point, right? The point is this: you love engineering and you love your family and you want to be able to enjoy and be successful in both areas of your life; and I believe you can.

Engineer Your Own Success: 7 Key Elements to Creating an Extraordinary Engineering Career,
First Edition. Anthony Fasano.
© 2015 The Institute of Electrical and Electronics Engineers, Inc. Published 2015 by John Wiley & Sons, Inc.

> You love engineering. You love your family. You want to be able to enjoy and be successful in both areas of your life. I believe you can.

To make it even more challenging, here I am, presenting you with all of the strategies in this book to supercharge your career and potentially make it even harder to achieve balance. On a positive note, I have achieved a good level of balance in my life, running a successful business, and raising three young children. It wasn't easy; I have studied all of the leading productivity experts in recent years from David Allen to Tim Ferris to Leo Babauta, and they have helped me to distill time management into three main areas of focus, which I will break down for you in this chapter. I will also give you real examples and tools that I found effective in managing my time.

The three areas that you must master to achieve balance between your career and life:

1. Be organized in all of your efforts.
2. Stay focused and productive at all times.
3. Avoid stress and worry at all costs.

If you can follow these three rules, you will put yourself in a position to create a well-balanced, low-stress, fulfilling life, both personally and professionally. How do you think I am able to run a successful business, maintain a job as an executive for an engineering association, maintain a podcast for engineers, write books, keep a successful marriage, and raise three young kids? Yes, these rules work if you follow them. These techniques work, and you can apply them, moving toward the oh-so-coveted feeling of balance in your life.

8.2   Rule #1: Be Organized in All of Your Efforts

Being organized impacts you in more ways than you think. Many professionals struggle with staying organized in the workplace. Between paperwork, e-mails, voice-mails, and meetings, there are so many different things to manage—we sometimes feel overwhelmed. Taking control of all of these items by getting organized can have a dramatically positive impact on your career and life on so many levels.

Professionals that are organized in the workplace typically lead less stressful and more productive careers and lives. In fact, researchers at the Princeton University Neuroscience Institute discovered that when your environment is cluttered, the chaos restricts your ability to focus [2]. When you get physically organized (your office, desk, even your calendar and appointments), it helps you to get organized mentally and alleviate the clutter in your mind that may be creating stress for you.

For example, have you ever had one of those days when there are a million things on your mind that you have to do, but you're not exactly sure where to start? You end up getting really stressed, going around in circles, and not being very productive. That clutter in your head creates an "I'm so overwhelmed!" feeling, which is definitely a productivity killer.

When you feel overwhelmed, your stress levels go through the roof, and your health can even be negatively affected. On top of all that, you may take the workday stress home with you, which in turn will affect your family life. This is something that I started paying attention to after reading Allen's best-selling book, *Getting Things Done: The Art of Stress-Free Productivity* [3]. Many of the thoughts in this section are results of me reading that book and then applying the strategies in an engineering workplace setting or helping other engineers to do so.

By becoming organized through creating to-do lists and having a structured calendar and a contact database, among other things, you will allay those "I'm so overwhelmed!" thoughts, and you will feel in control and on top of your game. This is when your stress level will be lowest and your productivity highest. When you feel in control, you will have a very clear frame of mind, and you will be able to focus on specific tasks.

The strategies in this section will help you to start getting more organized immediately in your day-to-day activities. I understand that this process may take time to set in place, and time is a very valuable commodity in our type of career. However, if you follow the recommendations made in this section, your increase in productivity will give you more free time.

8.2.1 Deploy a Minimalist Mind-Set

Rule #1 deals more with staying physically organized in the workplace. The biggest challenge to being physically organized is that in today's world, we have so many different things to organize or manage. We have our desks and offices at our place of work, and many people have home offices, multiple devices and inboxes (phones, digital notebooks, laptops, etc.), paper and digital filing systems, and so on and so forth.

The first rule covers one step that you can take toward getting more organized: adopt a minimalistic mind-set and eliminate as many things as possible in the workplace or in your arsenal of tools, applications, and programs you utilize. Another author that has helped me in this arena is Richard Koch who has authored several books on the Pareto Principle also known as the 80/20 Rule, or the Law of the Vital Few. This is a principle that was proven by an Italian scientist, Vilfredo Pareto, in 1906, when he observed that 80% of the land in Italy was owned by 20% of the people. Through studies, he proved that this ratio was very close to 80/20 in all walks of life [4]. Here are some examples as to how they might apply in the engineering world or in your specific day-to-day efforts:

- Of your company's income, 80% probably comes from roughly 20% of your clients.
- Of the finished products you produce (plans/reports/specifications), 20% of them are probably created during 80% of your time.
- You probably wear 20% of the clothes you own 80% of the time.
- You probably use 20% of the computer programs you own 80% of the time.

I think you get the picture. The key thing here is to try and recognize the 20% of things or efforts that are producing the 80% of results in your life and do more of them and less of the other stuff.

Here are two real-life examples I have implemented, one personal and one in my career. I live in the northeastern United States, so every year in the fall, I would have to take out my winter clothes and replace some of my summer clothes in my dresser and my closet. After reading one of Koch's books, I realized that I do wear 20% of my clothes 80% of the time, regardless of the time of year. Therefore, I spent a few hours one day eliminating some of my clothes so that all of my clothes both warm and cold weather attire could fit into my dresser and closet. I now no longer need to do the summer/winter changeover and have yet to have a problem. In fact, any stress that I had around worrying about what to wear or where to find it is gone, making at least this part of my life much easier.

In my professional life, I applied the 80/20 rule to my e-mail. I spent a few days monitoring all of the e-mails that came through my inbox and realized that of the hundreds of e-mails I receive in 1 week, only about 20% were really important. I spent some time unsubscribing from e-mails and setting up a rigorous filtering system to ensure that only important ones get to my main inbox. I can now go through and empty my entire e-mail inbox each day in a single, well focused e-mail session, and it has helped to eliminate those impulses of always having to check my e-mails.

> I'm offering you a deal: you commit to sacrifice, roughly, a week's attention, to diligently cleaning up your e-mail inbox and filters. This will allow you to only have to check e-mail once a day. Would you take that deal? I bet you would.

So when it comes to organization, the most important thing to remember is to eliminate as much as possible. Apply the 80/20 rule wherever you can and start to eliminate unnecessary things and actions. Here are some specific organizational tools and strategies that you can implement in your career.

8.2.2 Use the Old (and New) Trusty Notepad

One very simple way to start getting organized is to keep a bound notepad with you at all times throughout the course of the day. This notepad will allow you to capture all of the information that you receive in one easy-to-find location. Whether you have to write down information received from a phone call, a phone number from an e-mail, a conversation in the office with someone about a project, or even notes from a meeting outside of the office, everything will be recorded in your notepad. I started doing this early on in my career, and the impact this one organizational tool had on me was instrumental to my success. That is a powerful statement to make about an ordinary bound notepad, but it's the truth.

Please note I keep saying "bound" notepad. It's important that the notepad have a permanent binding on it so that you don't lose any of the pages. Many engineers use engineering pads to take notes, but the pages rip off so easily that it is possible information can be scattered everywhere. I use a 7 in."× 10 in." notepad that opens horizontally, not vertically (Fig. 8.1). It has a nice hard cover and a sturdy binding ensuring that the pages are hard to remove and can't be easily lost.

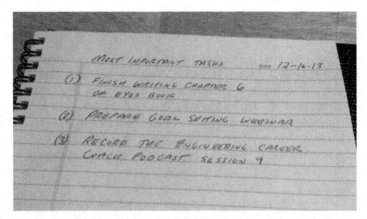

FIGURE 8.1 The trusty bound notepad. A bound notepad is one organizational tool that you can use to take down notes during your meeting.

Imagine that you are sitting at your desk and receive a phone call from a client. With your trusty notebook, you can easily jot down all of the information from the call, which you can review later if needed. If you have a meeting with your project team in the office to review some design documents, you can take notes from the meeting in your notepad and then generate a to-do list for that project later on. If you attend a project meeting out of the office, again take notes on your notepad and then prepare formal meeting minutes when you get back to the office. Once you run out of room in your notepad, you can simply label it with the date and store it on a bookshelf in your office and then go out and buy a new one to begin the process all over again. Labeling the pad with the date allows you to go back to it at any time and find information very easily, even from two or three years ago.

This notepad idea sounds so simple yet the value it will provide to you will amaze you. I have recommended this action step to many engineers, and when I see them a few weeks later, they tell me how much the notepad has helped them. Remember, though, that it is essential to transfer the collected information to the necessary proper sources as soon as possible. For example, if you write down a phone number on your pad, transfer that number into your contact database in your computer as soon as you can to avoid forgetting this step. If you take meeting minutes in your notepad, be sure to type them up in a Word document as soon as you get back to the office. You can even photocopy the page where you recorded your notes and give it to your administrative assistant to prepare the formal minutes. By doing this, you will guard against the worst-case scenario of losing or misplacing your notepad, which (unfortunately) I have also done in the past.

If you would prefer to use a digital notebook instead, there are many great tools out there like OneNote® and Evernote®. I use a combination of both paper and digital. The nice thing about the digital notebooks is that they can sync across any of your devices and they also allow you to put photos right into your notes, which can be invaluable for site visits and project meetings. However, I still find it hard to totally abandon the old-fashioned notebook because it can be very hard to take notes quickly digitally during an on-site meeting. Choose whichever combination makes sense to you, but be consistent.

I'm warning you right now: failure to utilize a bound notepad may subject you to something I call the "sticky note syndrome." This is when your desk and computer monitor are littered with bright-colored sticky notes, each one of them having a different phone number or e-mail address on them. Sticky notes are great if you want to try to lose valuable information. The one thing I believe they are good for is marking pages in books or design guides you are using. Other than that, please avoid using them at all costs to try and keep track of all the information you process during your workday.

> Warning! Failure to utilize a bound notepad may subject you to "sticky note syndrome," where your desk and computer monitor are littered with bright-colored sticky notes. This is not efficient or professional!

Start using a bound notepad today and all of your information will be at your fingertips whenever you need it.

8.2.3   Manage the Never-Ending Pile of Business Cards

I discussed in Chapter 7 the importance of building relationships throughout your career. In doing so, you will accumulate a ridiculous number of business cards (if you haven't already). These cards can really build up and be another contributing factor to that "I'm overwhelmed" feeling. Before you know it, you will have business cards everywhere: on your desk, in your desk drawers, in your briefcase, in your pockets, and so on and so forth.

It's essential that you have a system for dealing with the cards. I recommend that you have one digital database that you use to store all of your contacts. If you use Microsoft Outlook®, which most engineering companies do, store your contacts and relevant information in the Contacts module of the program.

> Remember: one business card could bring hundreds of opportunities to you.

As you collect business cards, it's important to transfer the data into your database as soon as possible. If you don't, not only will the cards start to pile up on your desk, but there is also a chance you will lose some of them in the process, which could be a costly mistake. Remember: one business card could bring hundreds of opportunities to you. I would recommend importing business cards into your database at least once a week. You may even be able to delegate that task to an administrative assistant. This would be the best-case scenario from a time management perspective. Once the data is imported, you can either dispose of the business cards (which for some reason people seem terrified to do) or, if it makes you more comfortable, file them away in a proper business card binder. Remember to apply the 80/20 rule. If the cards are in your digital database, will you really look at them again?

In Chapter 7's coverage of networking, I mentioned that it was a good idea to jot down some notes on the back of a business card, such as the person's specialty or even personal information such as their favorite hobby, or vacation spot. Be sure to enter that information into the notes section of your digital database as well so it won't be lost. Keep in mind that if you are entering your contacts into a company-wide database, your personal notes will be visible to everyone. Be aware that these notes may be lost if you transfer contacts between two different e-mail programs (and this is the only reason you may want to keep the actual cards). However, most professionals utilize the same system throughout their careers, so if you can easily transfer the information on the business card into a digital database, then choose this method rather than creating an extra pile of paper.

You should consider utilizing categories as you import your contacts. Many e-mail programs allow you to categorize your contacts by type. For example, some categories might include consultants, government agencies, contractors, and family. This can be extremely helpful because it will save you time when searching for a person in your database. Perhaps you want to look for an architect to team up with on a certain project. You can simply open the folder marked "Architects" and go through your list until you find the person you're looking for. Time is a valuable commodity that should never be wasted, and it's the little things like this that help your career shift from good or even great to extraordinary.

Many companies have public company contact folders where they ask all employees to add their contacts. This is absolutely fine, but be sure to also keep a copy of your contacts in your own personal Outlook folder, so if you ever leave the company, you can easily take all of your contact information with you. If you do leave your company and wish to transfer your own personal contact list, G-mail® (accessed through Google) is extremely compatible with Outlook and allows you to easily import all of your Outlook contacts. Of course, if your new employer uses Outlook, you may be able to transfer your contacts directly to your new account. Be sure that if you do take your contacts with you from one employer to the next, you are not violating a noncompete or nondisclosure clause. Upon leaving, some companies may deny you access to your old work computer to do this; some may even prosecute someone for doing so, as it could be stated that you are stealing confidential information or violating a nondisclosure or proprietary information agreement.

> The people you know and the relationships you build are the lifeline of your career. Be sure to organize their contact information in an efficient and effective manner to make the most of these connections.

By being aware of how you enter and use contact information in your databases that are available to you via mobile devices and computers, it will ensure that you always have the information with you, as most people have their phones synchronized with their contact database. This is important because if you are out of the office and want to call someone you met two weeks ago, you just have to type the name into your phone and all of the information is there. Not only is this convenient, again it saves time and puts you in that all-important position to succeed.

8.2.4 Remember That Missed Appointments Equal Missed Opportunities

Now that you have the tools to be a networking machine and you are ready to organize all of the business cards that you will be collecting, the next step is going to be organizing your calendar. In Chapter 7, I stressed the importance of reaching out to your contacts and building relationships by setting up lunch meetings and attending other social gatherings where they are likely to be in attendance. If all goes well, those social gatherings could turn into project meetings.

As you grow professionally, you are going to have more and more appointments to keep track of. The last thing you want to do is miss an appointment or have someone miss one with you. To avoid doing so, simply follow these steps.

8.2.4.1 Use Your Calendar Religiously

Select a calendar format that works best for you. Again, most engineers use a Microsoft Outlook calendar, while others, such as small business owners or independent consultants, may use a web-based e-mail program. You want to be sure that your calendar will allow you to invite other people to meetings and to receive a yes or no response from them. You will also want to choose a format that will allow an administrative or virtual assistant to access your calendar as needed. If you are looking to use a web-based e-mail platform, Google calendar contains these features. If you are still using a handwritten calendar, I strongly recommend you abandon it for a digital calendar, which will drastically reduce your chances of missing an appointment.

As soon as you set up an appointment, be sure to enter it into your calendar. This is where many professionals falter because they create an appointment over the phone or while they are out at a meeting, and they don't transfer it into their calendar:

- If you are using a bound notepad as discussed earlier in this chapter, be sure to transfer the information from the notepad into your calendar as soon as possible.
- If you are out of the office and don't like accessing your calendar from your phone, send yourself an e-mail or text message reminding yourself to put the appointment on the calendar at a later time.

8.2.4.2 Fill in All Pertinent Information

When creating a calendar appointment, enter the information in such a way that the invitation gets e-mailed to all those who should be attending the appointment. Some people don't feel comfortable sending appointments via e-mail to their clients and that's their preference. I personally feel that it is courteous to invite them with the calendar feature; it even saves them the time of having to put it on their own calendar.

Also, be sure that when entering the appointment into your calendar to set up a reminder, it will pop up and remind you of the meeting. I usually like to set the reminder a day or two before, which leaves me plenty of time to prepare for the meeting. Then I hit the "snooze" feature to an hour before as a "last call" reminder.

Since other meeting attendees will get the same reminder if you invited them through your calendar, be sure to include in your request the location and even a brief description of the objective of the meeting. This is especially important if you are traveling from appointment to appointment. It is very convenient to be able to quickly find the meeting address and description in one place. Most calendar formats will also allow you to attach a file such as a portable document format (PDF) to the calendar appointment, which may be helpful if it is a dinner meeting with a formal invitation that you may want to refer back to at a later time.

8.2.4.3 Confirm All Meetings

Last but not least, confirm the appointment with all attendees the day before by sending a short e-mail reminder. This is especially important if you did not invite them to the meeting using an electronic calendar system. I recommend doing this for all meetings. Spending 2 minutes to send out an e-mail can save you hours of traveling to a meeting that someone may not show up for.

Following these guidelines will minimize your chances of missing an appointment or having someone miss one with you. Remember: missed appointments equal missed opportunities.

8.2.5 Avoid the "I Am Not Sure What Color My Desk Is" Syndrome

The center of all of your organizational efforts is your desk. That's right—that big wooden or metal object that you sit in front of all day. That's kind of your home base for most people unless you are working in the field. The organization and appearance of your desk directly relate to not only your productivity but your success as well.

If your desk is neatly organized, you will be able to find things quickly and move easily from task to task. Losing or misplacing things will cost you time, and having to constantly reorganize your desk will also cost you dearly. If this is something that you struggle with, try to organize your desk at the end of each day. Take at least 10 minutes to dispose of papers you don't need and file away those that you are finished with for the day. You may want to have an inbox or a few stacks of current paperwork on your desk, but don't go what I call "stack crazy."

While piling papers may look neater, it doesn't help productivity. By taking the time to organize your desk at the end of each day, when you come in the next morning, you can start to be productive immediately. This can literally save you 30–45 minutes in the morning, where you previously were trying to figure out where you left off and sifting through piles of often unwanted paperwork.

You should also consider cleaning out your desk drawers, monthly or at least every few months. Apply the 80/20 rule as I am certain that most of the things in your desk drawers sit there and are never actually used. Get rid of them, and retain only those items that you use on a regular basis.

In addition to affecting your efficiency, the appearance of your desk is a direct reflection on you. A messy, disorganized desk may leave your supervisor, staff, or clients questioning your capability as a professional. They may think, "Well, if this engineer cannot manage a desk properly, how will managing my engineering project go?" This can also impact your performance review and the amount of respect earned from coworkers and clients.

> In addition to affecting your efficiency, the appearance of your desk is a direct reflection on you. A messy, disorganized desk may leave your supervisor, staff, or clients questioning your capability as a professional.

As you progress in your career, younger professionals will look up to you as a possible mentor, and having a chronically disorganized desk is not a good example to set. As a manager, if your desk is disorganized, your staff members may perceive that it is acceptable for their desks to look exactly like yours.

While having a messy desk may sound like nothing more than a joke to many people, it can have a negative effect on your career. Odds are that if your desk is messy, the clutter will start to spread to other parts of your office and your life. Take the time to straighten up your desk at the end of each day. It will be time well spent and you may even finally discover what color your desk is.

8.2.6 Prepare for Your Annual Performance Review

Your annual performance review is one of the few times at work that you should focus only on *you* and your personal achievements. You must be organized in order to do so. This is the main reason I placed this section in this portion of the book. Regardless of what kind of form the company asks you to fill out, prepare a list of all of your accomplishments achieved over the past year. Try to describe how the accomplishments helped the company. For example, "I joined a networking group and brought in new work to the company through this activity." Compiling this list of achievements can be daunting if you wait until the end of the year to do it. Instead, keep a running list of items and review and add to it every few months or weeks.

Your employer may take these accomplishments and their subsequent impact into account when considering your salary increase or promotion. Submit your list of accomplishments far in advance of your review (at least 30 days and ideally 60) to give the company time to consider these items when preparing for your review. It is pointless to prepare a list of accomplishments and present it for the first time at your actual performance review. This gives the company no time prior to the review to consider the accomplishments on your list that they may not have been aware of.

> It is pointless to prepare a list of accomplishments and present it for the first time at your actual performance review. This gives the company no time prior to the review to consider the accomplishments.

Many professionals get frustrated and complain that they feel they should have received a better review. They claim that their companies did not acknowledge all of their accomplishments over the last year. Did they ever think that their companies might not have known about these accomplishments? Especially when employed by a larger company, you must make these things known directly to your supervisors. Like the old saying goes, "The squeaky wheel gets the oil."

During your review or at any time during the year, don't be afraid to ask what specific steps you should take to reach a certain position in the company. When I wanted to achieve associate partner at a reputable engineering firm, I asked my supervisor during my review what steps I should be taking to do so. He gave me the steps, I pursued them, and I achieved them; I then became the youngest engineer ever promoted to associate partner in the firm. Being proactive in your own career is the best thing you can do for yourself.

When preparing your list of accomplishments for you review, consider including the following:

- Hours worked versus any goal the company may have set for you
- Seminars and training you attended and how the training improved your skills
- Groups or associations you joined or have belonged to
- Events you attended (this list may be long, but the longer the better)
- Accomplishments (i.e., an award, being elected to serve as a board member of a society, etc.)
- Intercompany committees that you may serve on
- Statistics of your staff (i.e., chargeability)
- Quantifiable measurements of the work you have brought in, using a dollar amount (this will assign a hard value to what you have accomplished)
- Amount you have collected in billings
- Goals for the upcoming year, as well as a progress report on goals you had targeted in your previous review

If your company does not give regular performance reviews, you should absolutely request one. When doing so, explain to your supervisor that you would like to have consistent reviews (every 6 or 12 months) to ensure that you are progressing and that your efforts are best serving the company. How can you track your progress as an engineer or know where you stand in the company without regular measurements?

> If your company does not give regular performance reviews, you should absolutely request one.

Remember: your annual performance review is your time to shine. It shouldn't be something you prepare for a day or so in advance and simply walk into the meeting hoping to get a typical salary increase. It's your chance to show the company what you are worth and to get what you deserve.

8.3   Rule #2: Stay Focused and Productive

You can be as physically organized as possible, but if you can stay mentally focused on the task at hand, your productivity, and subsequently time management, will be anything but balanced. I recognized this a few years ago and have sought out books and blogs to help me to maintain my focus and increase productivity. In doing so, I stumbled upon the Zen Habits blog (http://zenhabits.net) written by Leo Babauta through which Babauta writes articles about adopting a more Zen approach in life and at work to increase focus, productivity, and overall happiness.

> Your ability to focus will directly impact how productive you are in every sector of your work and life.

While Zen can be thought of in a religious sense, Babauta is referring to Zen in terms of a total state of focus that incorporates a total togetherness of body and mind. In fact, Babauta also read the book I referred to earlier, *Getting Things Done* (GTD) [3]. He loved it, but thought the processes in the book were cumbersome to implement, so he wrote a book, which is intended to be a Zen version of the GTD system. The book is called *The Power of Less: The Fine Art of Limiting Yourself to the Essential in Business and Life* [5], and I highly recommend it to all engineers. Many of the practices in this section were based on me taking Babauta's recommendations and implementing them into an engineering setting or helping motivated engineers to do so.

The following sections under this rule will provide strategies that you can implement to maintain focus in a world of distractions. Some of them will be more difficult to implement than others, but the steps for implementation are provided in this chapter. Please use them. Your ability to focus will directly impact how productive you and your team will be. Your productivity level will dictate the number of hours you will have to work to meet your project deadlines and achieve your career goals. The number of hours in one day will never change, but I have found that amount of work you get done in a day is directly related to your ability to focus.

8.3.1   Create Consistency Through Routines

In studying productivity over the years, I have found that one of the most critical factors in staying focused and productive is to implement a regular weekly routine that you follow consistently. For example, I have a working routine/schedule that is broken down on my calendar. It breaks up my days into four quadrants: early morning (before breakfast), morning (between breakfast and lunch), afternoon (between lunch and 5 p.m.), and evening (8:30 p.m. to 10:30 p.m.). Then on this schedule, I have tasks down in these quadrants that I do consistently during these times every week. For example, I typically do writing tasks like blog posts early every Monday morning and then finish them off in the morning slot. I usually record my podcast, The Engineering Career Coach, every Tuesday morning. I think you get the picture. See my actual calendar in Figure 8.2.

Many people think that working this way is robotic and stunts creativity. In fact, I believe the opposite is true. By doing regular tasks consistently at the same time each week,

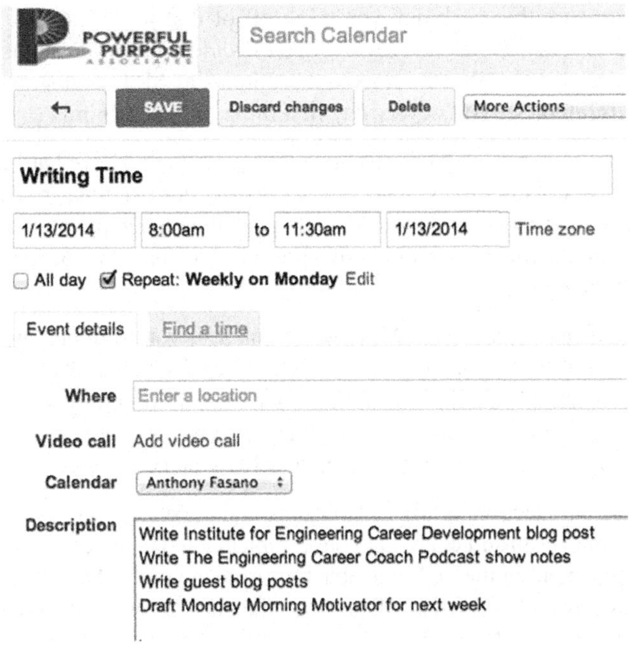

FIGURE 8.2 Sample calendar. Making a routine for yourself provides calm in the storm and allows creative thinking to flourish.

I know they are going to get done, which reduces stress. I also know when they are going to get done and I designate other times for creative work like writing, product development, and career coaching. I have experimented with this on and off in my career, and whenever I try to wing it and not have specific routines for accomplishing tasks, things get missed, which leads to poor results and a good amount of stress. One example that you may be familiar with is your timesheet. How much easier would your life be if you did your timesheet every Friday at the end of the workday instead of Monday morning? The answer is much easier, but we both know that probably very rarely happens.

Doing things in accordance with a routine also increases productivity because you spend less time planning or thinking about what to do. On a normal week, I usually write three blog posts, my Monday Morning Motivator, record a podcast session, hold a few coaching sessions, and keep up with e-mail and social media, all by 9:30 a.m. on Tuesday— yes, Tuesday. If you are looking for ways to increase your focus and productivity, consider implementing a regular weekly routine into your career immediately.

8.3.2 Establish Your Most Important Tasks Early Each Day

The next topic I would like to discuss is an idea I picked up from Babauta's book *The Power of Less* [5], most important tasks (MITs). You can't stay focused or productive if you are not clear on what tasks or goals are most important on a daily basis. I am sure you have a to-do list of some sort, which we will discuss in detail later in this chapter.

I am also pretty sure that each day you work off of that to-do list. The question that I have for you is this: "When and how do you decide what to-do items are the most important on your list?" Babauta refers to these items as MITs and stresses the importance of identifying these MITs every day first thing in the morning.

To identify them, review your to-do list and identify tasks that have hard deadlines and/ or will deliver BIG results for you or your company. For example, some MITs might include a proposal or project submission with a deadline, an application or a portion of one like the PE application, preparation of a presentation that must be given on a certain date, your timesheet, etc.

Another way you can think of your MITs is to think as if you were only going to achieve those three tasks for the day, and if you were able to achieve more, that would be a bonus. Once you have identified your three MITs, highlight them on your to-do list. I have actually found it easier to identify the day or evening before so when I get into the office the next morning I am focused on exactly what I want to achieve and can start moving right away.

Please don't take this section lightly. Establishing and pursuing MITs every day has been one of the most impactful strategies that I have implemented in my career and have helped engineers to implement in theirs. Being able to prioritize your tasks will ensure that your focus is on those 20% of tasks that will generate the 80% of results in your career.

> Being able to prioritize your tasks will ensure that your focus is on those 20% of tasks that will generate the 80% of results in your career.

8.3.3   Complete or Assign Your MITs First Thing Each Day

So, it's important to establish MITs every morning. I know many professionals that do this, but then once the workday starts, they abandon their MITs and work on whatever pops up during the day. As an engineer, there will always be "emergencies" that have to be dealt with, but engineers have a habit of making everything that arises an emergency, which is often not the case.

In order for this strategy to work, you must hold these MITs as sacred and make it a habit to stick to them until they are completed. For this reason, I have implemented early work time in my routine, which usually runs from 5:30 a.m. to 6:30 a.m. In this time period, I can usually accomplish two of my three MITs.

The other thing to consider with respect to completing your MITs is that you don't have to do them yourself. If you have a team of people that work with or for you, you may be able to assign one or some of them to the team. I often assign at least one of my daily MITs to one of my virtual assistants through a task management system. (I hire independent contractors who work for me on an hourly basis through sites like www. odesk.com or www.elance.com.) Having similar help will give you some more flexibility to handle emergencies that arrive while still staying on track to accomplish your MITs.

Identifying your three MITs each day and putting all of your energy into achieving them as early as possible in the day will dramatically increase your productivity—maybe not in the amount of tasks you complete, but certainly in the importance and overall impact of the results generated.

8.3.4 Control Your Own Schedule by Breaking Bad E-Mail Habits

E-mail is the most common form of communication in the corporate world today; however, it is also one of the biggest time wasters and distractions around. If you let it, constantly checking, composing, or answering e-mails will eat up so much of your time that your focus and productivity level will be drastically reduced. I once read a quote by Brendon Burchard, author of *The Millionaire Messenger* [6], saying that e-mail is someone else's way of managing your day, because by answering e-mails as they come in, you are constantly putting other people's needs before yours.

Following are some recommendations for effectively managing your e-mail. There are books dedicated to e-mail management alone, and one that inspired me to take control of my e-mail was one called *The Four Hour Work Week: Escape 9-5, Live Anywhere, and Join the New Rich* by Timothy Ferriss [7]. Ferriss inspired many of the following recommendations, which I have implemented and I am truly grateful for them.

Do not, I repeat *do not*, check your e-mail messages as soon as they arrive in your inbox. Doing so will continually interrupt your focus on whatever else you are trying to accomplish. If you get into this habit, you will literally be checking e-mails all day long. If your e-mail program has a notification that pops up or makes a sound when you receive a new message, turn it off.

Actually, stop reading this book right now, and turn off your e-mail notifications.

The same goes for your smartphone. Turn off the ringer or vibration that accompanies the arrival of new e-mails.

There are always some professionals who say that they have to check their e-mail as it comes in because they are working on time-critical projects, such as one under construction, where an e-mail could contain critical time-sensitive information. If that's the case, try pushing people toward calling you for emergencies instead of e-mailing you, maybe by using an away message initially on your e-mail explaining that you only check your account periodically throughout the day. However, state that you can be reached on your cell phone in the case of an emergency and provide the number. This won't work for everyone, but if you can get away with it, it will have a notable positive impact on your ability to focus on tasks for longer periods of time.

> Avoid checking e-mails constantly. If your e-mail program has a notification that pops up or makes a sound when you receive a new message, turn it off.

Set aside a few specific times during the day to check your e-mail. This will allow you to stay focused on your tasks in between these designated times. I know this is not an easy thing to do, but if you master this one technique, you will feel like you have taken back control of your life from the black hole of e-mail. Knowing in your mind that you have specific times for checking your e-mail will help you to forget about e-mail in between those time frames. Please at least try this system. The times of the day that you select to check e-mail will be totally dependent on your schedule, although I highly recommend not checking your account first thing in the morning if you can avoid it. For most people, mornings are the most productive time of the day, so the last thing you want to do is waste that time sifting through e-mails.

Try the following experiment over the next two business days. The first day, come into your office, and start checking and responding to e-mails immediately. The next day, start off by commencing work straight away. Start going through your "to-do" list, attacking things you want to accomplish that day. If your list includes e-mailing people, open a Word document, type the communication into the document, and then transfer them into e-mails later during your specified e-mail times. I'll tell you right now that on the first day you will not even start working on regular work-related projects until 9:30 or 10:00 a.m., and that is if your e-mail load is light. Yet, on the second day, you'll find that by not starting off buried in your e-mail account, by noon you will most likely have completed at least half of your MITs.

When you do spend time going through your e-mails, do it effectively, attempting to remove as many of them from your inbox as possible during your allotted time. Letting them pile up in your inbox contributes to those "I'm so overwhelmed" thoughts that I talked about earlier in this chapter. Along with those feelings comes stress and added pressure that will only weigh you down.

As you go through your e-mails:

- Try to delete them, save them to the proper location, or forward them to whomever may be handling that specific issue.
- Set up folders where you can store e-mails for a certain project or topic to be accessed at a later time. A particular message may not be applicable now, but may be needed down the road. Keeping these kinds of "not-pressing" e-mails in your inbox will just bog you down. In David Allen's book, *Getting Things Done* [3], he recommends a two-minute rule for handling e-mail, which basically says that if you open a message and can deal with it in two minutes or less, do it and be done with it (more on this in Section 8.4.2). He provides two other options: delegate it or delete it. This is a simple but very effective rule that I have successfully followed for years.
- Deploy filters on your e-mail program to prioritize messages as they come in. For example, you may have a folder set up so that all e-mails associated with a certain project or client automatically get directed into a particular folder. Alternatively, you can create folders for less important e-mails like newsletters so they skip your inbox entirely, and you can read them later when you have some downtime. Don't go folder crazy though; keep it as simple as possible.

I am not trying to degrade e-mail at all. In fact, if used properly, I believe it is one of the most powerful forms of communication available today. I just want to urge you to be smart in using e-mail and not to be a slave to it—use it in a way that is productive, not overwhelming.

8.3.5 Slow Things Down through Meditation

We live in a world of information overload where our mind is subject to a constant flow of nonstop information from articles, blogs, e-mails, television, podcasts, seminars, webinars, etc. This overload causes our minds to race, just trying to keep up with all of this information but also makes it very difficult for our mind to focus on one thing at a time. Have you noticed that? It's not even so much the physical interruptions, which I will discuss more in the next section, but it's the fact that our mind has so many different thoughts running through it that we just can't focus it.

> "Restore your attention or bring it to a new level by dramatically slowing down whatever you're doing." —Sharon Salzberg [10]

My research over the past few years on this topic has led me to the practice of meditation and reflection on a daily basis. Meditation is defined and performed in different ways, but I like to describe it as a process used to slow down or empty your mind of all of its thoughts. To achieve this state, I typically sit in a quiet place in my home office every morning with some soft music on and try to let all of my thoughts go. I have been able to do this for anywhere from 10 minutes to 30 minutes per day. Meditating consistently can dramatically increase your ability to focus throughout the day. Mediation also has been proven through studies [8, 9] to reduce stress reduction and facilitate better sleep patterns.

If meditation seems difficult or impractical for you, then try instead to implement some times of reflection into your day. Spend 5 minutes at your desk or on your couch, simply sitting still and quietly reflecting on your goals for the day or the things you have accomplished. This quiet time will give your mind a break from the fast-paced environment that it is typically subject to.

I once thought of mediation as an unattainable goal for someone as motivated and driven as myself; however, thanks to blogs like Zen Habits, I have successfully implemented meditation into my daily routine; and to be honest with you, I don't know how I would or did function without it.

8.3.6 Focus Intently on What You Are Doing

I have discussed several different strategies so far in this chapter that can help you to focus more and increase your productivity; however, there are two key things you must do to be able to implement these strategies.

First, you must put yourself in a position to accomplish these things, physically. For example, you can't check e-mails only during certain times per day if your e-mails are popping up in your face, and you can't focus on writing a report if your cell phone ringer is on high volume sitting right on your desk. Set yourself up to be able to focus and complete these tasks. Put yourself in an atmosphere where you can be successful. Most of this book was written on an airplane because there are very few distractions on an airplane. There is no Internet (unless you purchase it) and there are no phone calls or e-mails to interrupt you. I recognized the airplane opportunity, and I try to duplicate this atmosphere whenever I need to get things done. So I might go to a coffee shop that I know doesn't have Wi-Fi service available. You must consider this when deciding where or how you are going to work.

The second thing you must do to implement these strategies is you *must* do them consistently until they become habits or else you will abandon them. If you search online, you will see many articles saying that it takes 21–28 days consecutively of doing something to cement it in your brain as a new habit. However, the only official study that I could find, from the *European Journal of Social Psychology*, indicates that it takes approximately 66 days to ingrain a new habit into your brain [11]. At 66 consecutive days of activity, that habit is going to be as much of a habit as it is ever going to be. In other words, the habit has become automatic and that action associated with is never going to get more habitual. Whether it is 21 or 66 days, the bottom line is you have to do it continuously for it to stick.

> To implement these strategies, you must do them consistently until they become habits or else you will abandon them.

So, remember in order to increase your focus and productivity, be mindful of the atmosphere you subject yourself to and also practice new habits consistently until they become ingrained.

8.4   Rule #3: Avoid Stress and Worry at All Costs

Stress and worry are two feelings that can overwhelm anyone, engineers notwithstanding. That being said, engineers exhibit large amounts of stress due to unrealistic project deadlines often set by aggressive clients and pressure to constantly be billable from their superiors. *Actually, the career of an engineer is set up perfectly to be a stressful one, unless you take steps to alleviate stress in your career and life.*

Much of the stress that people experience is due to worrying that they do about everything. Will I finish this project on time? Will I pass this exam? Will I get accepted into this program? What if I don't? I heard a great quote when I was younger that I have kept at the forefront of my mind to fend off those worrisome feelings, whenever they creep in.

> Worrying is like a rocking chair. It gives you something to do, but gets you nowhere.
> —Glen Turner   [12]

This quote has helped me to pretty much eliminate worrying in my life, probably to my detriment at times. I never seem to worry about things, no matter how important they are. The bottom line is that worrying won't get you anywhere in life, so why do it? We often worry about things that we have absolutely no control over. Why? Worrying leads to stress and stress leads to unhappiness both personally and professionally. Here are some ways to reduce your stress that have worked for me, and many of my clients, in the past.

8.4.1 Simplification Through Elimination

This section addresses taking on a minimalist mind-set and eliminating physical clutter in your life to help you to get more organized. In this section, I am going to focus on eliminating mental clutter, a monumental challenge in the information overload era that we live in.

Think about this for a minute. How many different notifications or messages do you receive in one day and in how many different formats? Review the list below and estimate how many of each of these you are subject to on a daily basis:

- Phone calls on your home, office, or cell phone
- Text messages on your cell phone or computer
- E-mails from your work or personal accounts on your phone, tablet, computer, etc.
- Messages through social media sites like Facebook, LinkedIn, and Twitter

I could come up with more, but I think you get the picture. It's so easy to get distracted and stressed out dealing with all of these different notifications, which is why I recommend eliminating some.

For example, I have an iPhone®, an iPad®, and a laptop. However, I only use the devices for specific tasks. I purchased the iPad in an effort to eliminate paper books from my office and now use it primarily for reading books and articles. When I first purchased it, I was checking e-mail on my phone, then the iPad, and then the laptop. It was so stressful and mentally tiring—it was like I could never get away and focus on anything.

Select a few technological devices, programs, and applications that best serve you and eliminate the rest. Take the time to really learn the ones that you select and use them to stay productive. Less is more, especially when you are trying to reduce stress in your life.

Less is more, especially when you are trying to reduce stress in your life.

8.4.2 Empty Your E-Mail Inbox Twice Per Day

Okay, so you read the title of this section and started laughing, right? Well, believe it or not, this is no joke. I highly recommend that you empty your e-mail inbox twice per day. This is something I tried after reading Babauta's book *The Power of Less* [5], and it has definitely been life changing.

The reason that I feel this is so important is because I have noticed that the amount of stress that people experience is directly related to the number of e-mails in their inbox. What is the connection? People think of e-mails as "things to do," so the more e-mail they have, the more things they have to do, the more things to worry about and stress over. You see where I am going with this. It is for this reason that it is important to clear your e-mail inbox on a very regular basis—ideally daily. The big question, of course, is, how to accomplish such a feat? I will lay out the answer in three steps:

1. Aggressively filter your e-mails so that only important messages are getting into your primary inbox. I talked about this in Section 8.3.4, but it's worth mentioning

again. Every time you go through your primary inbox and see an e-mail that is not important, either unsubscribe from the sender or filter e-mails of this nature out so they skip your inbox and go to a folder that you can check less frequently. Failure to do this will make it virtually impossible to empty your inbox and subsequently difficult to reduce stress in your life.

2. You must do your best to go through your inbox completely, consistently, once or twice per day. I am not talking about glancing at your e-mails—I am talking about a full e-mail session where you go through your entire inbox. If you do not do this, then your e-mail will pile up to the point where you won't be able to catch up. At one of my Engineer Your Own Success seminars, an attendee told me that he had about 4000 e-mails in his inbox from the past few years. He asked me for advice on what to do, and I said delete them all or filter them all to a folder outside of your inbox immediately. He seemed taken aback, so I asked him if he had looked at any of the e-mails older than one month old recently. His response was, "No." Filtering and checking your e-mail regularly will help you to avoid a 4000 e-mail pile up as well as the stress that comes along with it.

3. Alter the way you check your e-mails. This process is from Allen's book *Getting Things Done* [3]. Allen says when checking e-mail, you should take a 4-D approach: delete, do, delegate, or defer. If the e-mail is junk or no longer needed, delete it. If the e-mail is something you can do in two minutes or less, do it and delete it or save in another location. If you cannot accomplish the task in two minutes or less, delegate the task to someone else, or add it to you to-do list to be done at a later time. In all four scenarios, the end result should be that e-mail no longer haunts your inbox.

I hope you now take my advice of emptying your inbox much more seriously. Remember rigorous filtering and consistent focused sessions of doing, deleting, delegating, or deferring will greatly reduce the number of e-mails in your inbox and the stress associated with them.

8.4.3   A Good To-Do List Can Work Wonders

Whether it's a document on the computer or scribbles on a napkin, we all have a to-do list. Some of us have multiple lists, maybe one for each project or aspect of our jobs. If you do not currently use a to-do list, please consider creating one, as a good one can really guide you through a highly productive and low-stressed day and ultimately make a difference in your career. Many professionals start using a to-do list, which ends up lasting a few months, before they lose it or just get away from updating it regularly because they become overwhelmed by the length of it. Based on my experience, I have found the following guidelines useful for effectively utilizing a to-do list.

First, decide on using either a written list or one composed on the computer. I recommend the computer version because it is easy to update and harder to lose, but you can make the decision based on your comfort level. A spreadsheet format has always worked best for me. Organize your tasks by categories, which might include different projects, business development activities, professional association items, general tasks, or training. I have recently tried a few web-based to-do lists like asana.com, which I really like, but they are web based, meaning that if the Internet or their site goes down, you lose access. So I now use a combination of a web-based to-do list, a good old Excel spreadsheet, and my calendar. Of course, many of the notes that I scribble down in my trusty notepad get transferred into my to-do list at the end of each day.

Go through each category and add your to-do items under each one. For one project, you might have items such as *finish report*, *call client*, and *prepare for meeting*. You shouldn't have to list meetings on your to-do list because if you followed the advice provided in Section 8.2.4, all of your appointments are on your calendar with reminders. It's important that beneath each category you really break out the different tasks as discussed here, because if the tasks are too generalized, you may not know what to do next when glancing at your list. You might also want to utilize Outlook for your to-do list; however, I have found that a spreadsheet or a to-do list application is easier to customize to my liking.

Once you have all of your tasks listed, I recommend you review them at the beginning or end of each day (preferably the end) and determine your MITs for the next day as discussed in Section 8.3.2. You can color-code your MITs and other tasks if you'd like, so perhaps use yellow highlights to indicate your MITs. The color pink might be used to denote secondary tasks that you can tackle when you finish the yellow ones. As you progress in your career, you may assign other color codes to tasks that you have delegated to certain staff. If you have a task that you are waiting for information on, you can color-code that as well. For example, if I have to finalize a report and I am waiting for information from someone, I will highlight that task gray until I get the information. I am not trying to make you go color crazy here, but by using this system, you can get a pretty good handle on your workload by just glancing at your list. It also allows you to check periodically throughout the day to see if you are on target to meet your daily objectives, and if you are not, you can either enlist help from others or modify your goals.

> Regardless of the kind of to-do list you utilize, please do not adopt the belief that you have to complete the list daily to be productive or successful.

Many people claim that they do not use to-do lists because too many items come up during the day that are not planned, so they fear that they would be constantly adding things to their lists. In my opinion, this is a poor excuse. The beauty of the to-do list is the ability to easily add and remove tasks and juggle your workload. As I mentioned earlier in this chapter, if you rely on committing too much information to memory, you are going to stress yourself out and increase the risk of failing to meet deadlines. For me, no matter how many tasks I have to do, as long as I can see them all laid out in front of me, I don't stress out about them.

Regardless of the kind of to-do list you utilize, please do not adopt the belief that you have to complete the list daily to be productive or successful. Unfortunately (or fortunately, depending on how you look at it), your to-do list will always have items on it. It has to. It's a list and if it didn't, you might not have a job. Just consider it a guide to help you to be productive throughout the course of your day. Enjoy your new to-do list and the increased productivity that will come along with it.

8.4.4 Keep Your Body (and Mind) in Shape

Most people today have so much going on in their careers and lives that they often don't have or make the time for exercise. Before I had kids, I used to go to the gym every morning before work. Now with three kids and my own business, the gym is a distant memory. However, I recently made a change in my life where I have incorporated practices to exercise both physically and mentally. This change has had a very positive impact on my life, especially with respect to stress reduction. I am in no way trying to push my practices upon you. In fact, it's imperative that you establish exercise routines that work for you. But I hope, at a minimum, this section provides you with inspiration to do so.

Every morning, I wake up at 5 a.m. and do two things: one for mental exercise and the other physical. I start off my day with a mediation session (see Section 8.3.5). This session lasts anywhere from 10 to 30 minutes. It allows me to clear my mind and start the day fresh with respect to thoughts in my mind. This has greatly improved my ability to focus throughout the day.

Immediately following my meditation session, I spend 10 to 30 minutes doing what is known as T'ai chi ch'uan. T'ai chi ch'uan, usually shortened to Tai Chi, is an internal Chinese martial art practiced for both its defense training and health benefits, and it involves two primary features. The first is a slow sequence of movements, which emphasizes a straight spine, abdominal breathing, and a natural range of motion. The second is different styles of pushing hands for training movement principles of the form with a partner and in a more practical manner. I discovered Tai Chi through research and purchased a DVD online, which I follow every morning.

Studies have shown that among other health benefits, Tai Chi improves balance control, flexibility, and cardiovascular fitness. The practice can increase psychological well-being including reduction of stress, anxiety, and depression and enhanced mood in community-dwelling healthy participants and in patients with chronic conditions [13]. I have experienced many of these benefits myself.

The other form of physical exercise that I perform each day is walking. I take a 30 to 40 minute walk every day. The walk is usually fast paced and done outdoors, as opposed to on a treadmill, weather permitting. This has probably had the biggest impact on my physical fitness. I read a book once called *The Monk Who Sold His Ferrari* by Robin Sharma [14]. It is a wonderful fable that describes a top trial lawyer's journey into the Himalayan Mountains where he discovers his true self while living with a group of wise sages. While the book is fiction, it talks about the physical fitness of these wise sages and claims that there main form of exercise was rigorous walking every day. This led me to research the benefits of walking.

I found through research that by simply walking every day, you could improve your overall physical and mental well-being dramatically. In one case study that I found, Italian researchers enlisted 749 people suffering from memory problems in a study and measured their walking and other moderate activities, such as yard work. At the four-year follow-up, they found that those who expended the most energy walking had a 27% lower risk of developing dementia than the people who expended the least. This could be the result of physical activity's role in increasing blood flow to the brain [15].

I usually try to walk every day at lunchtime or incorporate family time with walk time and walk my daughter to school, or take a walk with my wife. You can also increase the amount of time you walk during the day through your typical errands. For example, when I have to go to the store, I walk to the store if possible or park as far away from the store as I can to force myself to do some walking. At times, I will listen to podcasts when walking so I am improving both mind and body at the same time.

These are some of the exercises that I have implemented into my daily routine to help keep both my mind and body in shape. In order to implement these types of practices into your daily routine, you will have to create these new habits. In order to create new habits, you will have to engage in the same activity 21–66 days in a row (see Section 8.3.5). This is exactly how I formed my mediation, Tai Chi, and walking habits and you can start implementing your own exercise habits too. For a long time, I neglected working out because I didn't want to sacrifice working time to do so. However, I have since found out that these practices, while reducing my work time, have made the time that I work extremely more productive and enjoyable.

> Making time for exercise may reduce work time. However, the time I do spend working is now more productive and enjoyable.

8.4.5 Eat and Sleep Well

Your level of stress and overall energy is directly related to both your eating and sleeping habits. When people think about their careers, they seem to totally neglect the fact that these two critical life functions are directly related to your professional success. I learned this the hard way. I used to eat foods containing large amounts of starchy carbohydrates that were very difficult for my body to digest. This led to excess weight and poor sleeping habits because my body was busy digesting foods (or trying too) rather than getting a restful sleep. I am certainly not a doctor, and I am not offering medical advice in this book. I am simply explaining some practices that I have used in my life that have given me much more energy and focus.

With respect to eating, recently, I implemented a new diet where I have greatly reduced the number of starchy carbohydrates (cereals and breads) and increased my fruit and vegetable consumption. I really enjoy eating leafy green vegetables such as kale, spinach, chard, etc. Not only do these vegetables taste great, but studies have shown that green vegetables improve digestion and the overall strength of your immune system [16]. I can attest to the fact that I very rarely get sick.

Sleeping is another critical function of the body that many people discount as important. One of the key factors to getting a good night's sleep is the type of food that you eat. I have found that eating dinner as early as possible and then refraining from late snacks allows for my body to digest most of the food eaten, allowing for a more restful sleep. The types of food taken can affect digestion and, ultimately, sleep habits. I have also found, and this was a huge life changer for me, that going to sleep earlier and waking up earlier gives me more energy. I now try to go to bed by 10 p.m. (11 p.m. the latest) and wake up each morning at 5 a.m., following the early morning routine. I find that on most days by 6:45 a.m., I have already mediated, have done Tai Chi, and have completed one of my MITs.

As I said, I am not a specialist in these fields, but I wrote this section to highlight how changes in my own diet and sleep habits positively affected my career and personal development. I hope that through sharing my routine, it will inspire you to create a routine for yourself that fosters healthy eating and sleeping habits.

8.5   Work–Family Balance Is Achievable

I would say that about 75% of the engineers that I provide coaching to are not happy with the amount of time they get to spend with their family. The truth of the matter is this: most engineers that seek out career coaching are very motivated and driven in their career. That attitude usually produces a fast-tracked career where the engineer rapidly climbs the corporate ladder. A career of this type comes with many responsibilities and expectations that undoubtedly interfere with family life, mainly by forcing engineers to work long hours in the offices.

You can beat this typical routine, but you must be intentional about it and have a plan to do so. This final section of this chapter will lay out a few steps to help you to define what balance means to you and achieve it in your career.

8.5.1   Define Work–Family Balance

The first step to achieving anything, as I discussed in Chapter 1, is to define it. Many engineers will tell me they want more balance between work and family, but when I ask them to define what that balance means, they cannot do so. *The first step to achieving more balance in your career and life is to define it.* Does work–family balance mean you are only going to work a certain number of hours during the day? Does it mean you are only going to work a certain number of hours in the office? Does it mean you are going to have dinner three out of five nights of the week with your family? What does it mean to you?

> Many engineers will tell me they want more balance between work and family, but when I ask them, "Can you define what more balance means?," they typically answer, "No."

Once you have defined work–family balance, then you can use many of the strategies in this book to achieve your goals. Therefore, if you want to be home every night for dinner, maybe you need to wake up early and complete one of your MITs at home in the morning before you go to the office. If you want to make it to your kid's sporting events more regularly, maybe you need to delegate better (see Chapter 9).

I believe the strategies in this book can help you to achieve work–family balance, once you understand what it means to you. Work–family balance is very important to me. In fact, it is one of the main reasons I decided to start my own business. You may have realized by now that my day is built around family and personal time. Below is my typical daily schedule:

5:00 a.m.	Daily mediation
5:15 a.m.	Tai Chi
5:30 a.m.	Work on MITs/delegate to staff
6:30 a.m.	Breakfast/prep kids for school
8:00 a.m.	Work on MITs
11:00 a.m.	E-mail session/social media
12:00 p.m.	Exercise (usually 30 minute walk)
12:30 p.m.	Eat lunch
1:00 p.m.	Phone calls
1:30 p.m.	General work/coaching
4:30 p.m.	Review day/reorganize
5:00 p.m.	Dinner prep/family time
8:30 p.m.	Work (two nights per week)
10:00 p.m.	Reflection/journaling/tea
10:30 a.m.	Bed

8.5.2 Build Flexibility into Your Career

One of the most important aspects of achieving work–family balance is flexibility. When you build flexibility into your career, you are giving yourself more freedom as to how and where you spend your time. This is critical if you want to have the ability to spend time with your family on your own terms.

> Built-in work flexibility gives you more freedom to manage your time. This is critical if you want to have the ability to spend time with your family on your own terms.

Some actions you might take to increase flexibility in your career include but are not limited to the following:

- Negotiate the ability to work from home one or several days per week.
- Gain enough trust from your employer to be allowed to set your own hours.
- Train your staff to be able to cover for you when you are out of the office. (Don't worry; they won't take your job if you have developed all of these skills I discussed in this book.)
- Start your own company.
- Hire a virtual assistant to assist you with personal/professional tasks. This is not as expensive as you might think, as low as $5 per hour using sites like odesk.com and elance.com. The book *The Four Hour Work Week* by Ferriss [7] provides many strategies and step-by-step instructions for effectively utilizing virtual assistants.

These are just a few examples of things you can do to build flexibility in your career; however, in order to be able to implement some of these, you will need leverage. If you develop the skills discussed in this book, especially the ability to bring in new business to your company, you will have the leverage needed to negotiate many of these items into your career and life. I did, and I now can enjoy both my family and my career.

8.5.3 Be Present in the Moment

Whether or not you define work–family balance or build flexibility into your career, one thing that you can do immediately to improve work–family balance is to be focused and present in the moment on whatever you are working on at that time. For example, if you are at your son or daughter's soccer game, don't spend the whole game going through e-mails on your cell phone. Watch the game and be there with your child completely.

This is something that a good friend of mine, Christian Knutson, author of The Engineer Leader blog, has told me time and again. In fact, this is a message he sent me on the topic and I want to share it with you (used with permission):

> *I've come across a concept from a number of readings, which I now practice in solving the "balance" challenge: dynamic equanimity. The main concept is that balance doesn't exist except with objects that are at rest—i.e., a rock or being dead. Since I'm neither, I quickly absorbed the concept.*
>
> *What it means in practice is that at some moments you're focused on your work, at others you're focused on your hobbies/projects/pursuits, and at other times you're focused on your family, friends, etc. The bottom line is that you focus 100% on what you're doing at the moment and, more importantly, what needs to be attended to at the moment. (If you Google it, you'll come up with the Wikipedia definition of equanimity:* "**Equanimity** is a state of stability or composure arising from a deep awareness and acceptance of the present moment." *You'll also find that it's a central theme of Hinduism and a main focus of Yoga. If you can accept this, then good. I do and it works.)*
>
> *I've been approaching the "balance" matter for 2 years now with this mind-set, and it works for me. I don't worry, ruminate, or complain about my balance. Not surprisingly, my wife, my children, and my various work/art pursuits don't suffer either. I do what needs to be done.*
>
> *I'll wrap up with this quote from Osho's* "Book of Understanding" [17], *which I believe captures the concept of dynamic equanimity pretty well:* "Live life in all possible ways; don't choose one thing against the other, and don't try to be in the middle. Don't try to balance yourself—balance is not something that can be cultivated. Balance is something that comes out of experiencing all the dimensions of life. Balance is something that happens; it is not something that can be brought about through your efforts. If you bring it through your effort it will be false, forced."

I think Knutson's statement here truly sums up the message I have been trying to get across in this section. Understand what work–family balance means to you, give yourself some flexibility to achieve it, and always ensure that you are focused on what you are experiencing in the moment.

8.6 Key Points to Remember

1. The three key points to effectively utilizing your time include being organized in all of your efforts, staying focused and productive at all times, and avoiding stress and worry at all costs.

2. By becoming organized through creating to-do lists and having a structured calendar and a contact database among other things, you will allay those "I'm so overwhelmed!" thoughts, and you'll feel in control and on top of your game.

3. Try and recognize the 20% of things or efforts that are producing the 80% of results in your life and do more of them and less of the other stuff.

4. Keep a bound notepad with you at all times throughout the course of the day. This notepad will allow you to capture all of the information that you receive in one easy-to-find location.

5. As you collect business cards, be sure to transfer the data into a master database as soon as possible. One business card could bring hundreds of opportunities to you.

6. Keep your calendar up to date and make an effort to confirm all appointments. Remember: missed appointments equal missed opportunities.

7. Prepare for your annual review well in advance, and submit documentation to your supervisor, to give him or her time to review and consider your accomplishments that he or she may not have been aware of.

8. Your productivity level will dictate the number of hours you will have to work to meet your project deadlines and achieve your career goals.

9. Implementing a regular routine that you follow consistently daily or weekly can dramatically increase your focus and productivity because there is less planning required throughout your day and more doing.

10. Establish and pursue or delegate three tasks each day, which you designate as most important tasks (MITs). Attempt to complete or assign these MITs to others as early as possible during the day.

11. Take control of your day by setting up specific times to check your e-mail; otherwise, other people will be dictating your work schedule. This may be the most important point in this entire book.

12. A good to-do list can act as a guide to help you to be extremely productive throughout the course of your day.

13. Keeping your body and mind in shape through exercise, reflection, and meditation will positively impact both your productivity and enjoyment levels.

14. Your level of stress and overall energy is directly related to both your eating and sleeping habits. Be wary of this when establishing your diet and sleeping habits.

> ### *A Boost from Your Professional Partner*
>
> *To achieve work–family balance, you must first define exactly what it means to you. Once you define it, build flexibility into your career to help you achieve it, and always ensure that you are focused on what you are experiencing in the moment. Focus, produce, achieve.*
>
>
>
> Your Professional Partner,
> —Anthony
>

References

[1] J. Tichio, Greatest Inspirational Quotes: 365 Days to More Happiness, Success, Motivation, Scotts Valley, CA: CreateSpace, 2013, p. 91.

[2] S. McMains and S. Kastner, "Interactions of top-down and bottom-up mechanisms in human visual cortex," The Journal of Neuroscience: The Official Journal of the Society for Neuroscience, vol. 3, no. 2, pp. 587–597, January 2011.

[3] D. Allen, Getting Things Done: The Art of Stress-Free Productivity. New York: Penguin Books, 2002.

[4] V. Pareto, Manual of Political Economy. New York: A.M. Kelley, 1971.

[5] L. Babauta, The Power of Less: The Fine Art of Limiting Yourself to the Essential in Business and Life. New York: Hyperion, 2008.

[6] B. Burchard, The Millionaire Messenger. New York: Free Press, 2011.

[7] T. Ferriss, The Four Hour Work Week: Escape 9-5, Live Anywhere, and Join the New Rich. New York: Harmony Books, 2007.

[8] J. Ong and D. Sholtes, "A mindfulness-based approach to the treatment of insomnia," Journal of Clinical Psychology, vol. 66, no. 11, pp. 1175–1184, November 2010.

[9] C. R. Gross, et al., "Mindfulness-based stress reduction vs. pharmacotherapy for primary chronic insomnia: A pilot randomized controlled clinical trial," Explore (NY), vol. 7, no. 2, pp. 76–87, March–April 2011.

[10] S. Salzberg, Real Happiness: The Power of Meditation: A 28-Day Program. New York: Workman Pub., 2010.

[11] P. Lally, et al., "How are habits formed? Modelling habit formation in the real world," European Journal of Social Psychology, vol. 40, no. 6, pp. 998–1009, 2010.

[12] Guideposts, Wisdom Quote by Glenn Turner [Online]. Available at http://www.guideposts. org/inspirational-quotes/11001 (accessed on April 14, 2014).

[13] C. Wang, et al., "Tai chi on psychological well-being: Systematic review and meta-analysis," BMC Complementary and Alternative Medicine, vol. 10, p. 23, 2010.

[14] R. Sharma, The Monk Who Sold His Ferrari. New York, NY: HarperCollins, 1996.

[15] G. Ravaglia, et al., "Physical activity and dementia risk in the elderly," Neurology, vol. 70, no. 19, Part 2, pp. 1786–1794, May 2008.

[16] L. Rankin, et al., "The transcription factor T-Bet is essential for the development of Nkp46+ innate lymphocytes via the Notch pathway," Nature Immunology, vol.14, no. 4, pp. 389–395, 2013.

[17] Osho, The Book of Understanding: Creating Your Own Path to Freedom. New York: Harmony, 2006.

9

Be a Leader Every Day

Before you are a leader, success is all about growing yourself. When you become a leader, success is all about growing others.

—Jack Welch [1]

9.1 You Are a Leader

I would like to start this chapter off by convincing you that you are a leader. Many professionals think that they can only be considered a leader if they are the president of a company or if they are responsible for overseeing a certain number of individuals. That is what's called a limiting belief—a belief that you hold for no specific reason, which holds you back from reaching your maximum potential. I want you to let go of this limiting belief today and understand that you are a leader right now.

The dictionary defines a leader as a person or thing that leads. I like Bruce Schneider's definition of leadership in his book *Energy Leadership* [2]. Schneider defines a leader as someone who has the ability to inspire people to take action, including themselves. You move yourself into action each and every day, whether it's in your current position, in your job search, or even in your personal life. You motivated yourself to take action and read this book, which in itself tells me that you are, indeed, a leader.

Engineer Your Own Success: 7 Key Elements to Creating an Extraordinary Engineering Career, First Edition. Anthony Fasano.
© 2015 The Institute of Electrical and Electronics Engineers, Inc. Published 2015 by John Wiley & Sons, Inc.

I like Schneider's definition because it can be applied to everyone. Think about it. Presidents lead nations, executives lead companies, teachers lead students, and parents and caregivers lead their children, and right now whether you realize it or not, you are leading someone else through your own actions.

We are all leaders and we decide each and every day how we lead. In this chapter, I will provide you with positive ways to develop the leadership abilities that reside deep down inside of you, even if you haven't recognized them yet.

> We are all leaders and we decide each and every day how we lead.

9.2   The Power of Positivity

One very simple thing you can do to help yourself develop, as both a leader and a person, is to be positive. That's right—adopt a positive outlook in everything you do, say, and think. There is so much negativity in the world today, and I have seen firsthand how it has impacted many engineering professionals. A vast majority of the professionals that I work with say too often, "Every day is a battle" or "I am fighting my way through every day." You shouldn't have to fight or battle through your career; you should enjoy it. These negative feelings are caused by your perception. If you want to live a life of negativity, that is your choice. However, if you want to learn how to live life on the positive side, this section will help you to start to shift your attitude.

In addition to the impact that negativity has on the relationships in your life, studies have shown that positive people are generally healthier than negative ones [3]. This has to do with the amount of cortisol, a steroid hormone that gets released into the body in response to stress. A positive person typically produces less cortisol, which is important to your overall health, as cortisol suppresses the immune system. This alone is a good reason to become more positive [4]. In Appendix A, there are books on mind/body health, including *The Relaxation Response* by Benson and Klipper [5].

> When you spend time with negative people, you start to adopt their attitudes and even their beliefs; eventually, their negative perspectives will rub off on you.

The first thing you should consider doing is to surround yourself with positive people. When you spend time with negative people, you start to adopt their attitudes and even their beliefs; eventually, their behavior will seem normal to you. This is an important step to staying positive, but not always an easy one. There may be a coworker who constantly comes to talk with you or invites you out to lunch and rattles off one negative thought after the next. Trust me when I tell you to stay away from these people. If you have to work with them, do just that, and limit conversations with them outside of the regular office environment. As soon as you start surrounding yourself with positive people, you will start to see a big change in your overall perspective on life.

In addition to surrounding yourself with positive people, try doing at least one nice thing each day—something that usually wouldn't be a part of your daily routine. Maybe you could take a coworker to lunch just to catch up. At that lunch, you could offer yourself as a mentor. Perhaps you could send an e-mail to a friend or coworker thanking that person for past support. Doing something as simple as holding a door open for another person, or smiling at people when you pass them on the street, will cause your positivity meter to rise.

Start smiling more, stand up taller, and walk proudly. These physical movements will help you feel better about life in general and improve your self-confidence. These actions will change the way people look at you and even change the way they feel just by being around you. As a leader in the workplace, actions like this can affect an entire department in a positive way.

Last, but not least, use the Internet to help you stay positive. There are thousands of websites containing positive articles about personal development and leadership that can instantly shift your mood. I send out weekly inspirational e-mails to engineers, which you can sign up for at www.EngineeringCareerCoach.com. The same goes for reading positive books. There are so many good ones out there that can be found with a quick search; in fact, I have listed quite a few of these books in the Recommended Reading section at the end of this book.

Positivity is an important behavioral trait to develop as a leader because your attitude will affect those that you lead. I am not saying that negative people can't be successful leaders, because they can. However, negative leaders may be less healthy due to high levels of stress and poor-quality relationships [3]. You want your extraordinary career to be a positive, enjoyable one, and these strategies will help you obtain just that.

9.3 Great Leaders See Only Opportunity

Great leaders have an ability to see opportunity in every situation. You have probably been around people that focus mostly on the negative aspect of things or immediately identify what is wrong instead of first complimenting what is right. This mentality may be related to anxiety because a supervisor's job is on the line and in the hands of subordinates; however, it is a common approach. A design team typically exerts immense effort to finish a project, and unfortunately, many times, the only pieces that stand out are errors and mistakes. Let's work to change that toxic atmosphere.

In this section, I want to give you some strategies for always remaining positive and opportunistic in your career. I specifically want to share with you two questions that were shared with me when I read the book *Energy Leadership* [2] in conjunction with my training at the Institute for Professional Excellence in Coaching; they changed my entire outlook on life.

There are opportunities in every situation. However, sometimes, the opportunities are easy to see, and at other times, you will really have to look hard to find them. It's important to know that they are always there. For example, imagine that a problem arises on one of your projects where a certain material that you specified can't be used during production because the lead time to obtain it is too long. The manufacturer calls to tell you about this issue and you can choose how you react.

One way to react is to get upset. You might assume that there isn't any way to replace that particular material. Both of these thoughts fall into the category of a negative, close-minded approach.

On the other hand, you could ask yourself, "Where is the opportunity in this situation?" In responding to that question, maybe you realize that this situation will force you to find a new material to use that may give you more flexibility with this design in the future. Perhaps the new material will even be better than the original material specified, and you never would have discovered it if this issue had not surfaced.

Most people, when presented with the previous example, would focus on the negative, and that is unfortunate. However, you don't have to fall into such patterns. If you want to have an extraordinary career, you need to develop new habits that allow you freedom to explore possibilities. *You can decide today to start looking for opportunities in every situation by focusing on the positive and overlooking the negative.*

Let's look at another example. Over several months, you study very hard, take an examination for a license, and you do not pass the exam. Of course, there may be some immediate frustration on your part, which is totally understandable. Yet the way you react after that immediate frustration will define your success with this task moving forward.

If you take a negative approach and remain angry because you studied so hard and spent so much time preparing and you still didn't pass, it is going to be very hard to prepare again and retake the exam. If you ask yourself the second question, "How can I learn and grow from this situation?" you will force yourself to look for the positives or opportunities in the situation and use them as you move forward. By taking the exam and seeing the questions it contained, you may have realized that you need to make some changes in your study habits. These changes may not only help you to pass the exam the next time around, but they may also help you to become more at ease when taking exams further down the road. You have learned about yourself a bit and how you handle the challenges when confronted with what appeared to be a bad situation.

As an engineering professional, you will be faced with problems on a daily basis. For me, it works to call them "challenges," because the way you respond to them is critical to your success as an engineer, as a leader, and as a person. If you are always looking for growth and opportunity, you will use these situations to your advantage. On the flipside, if you are constantly harping on the negatives and asking yourself questions like, "Why does this always happen to me?" then you will feel like your career is stagnating.

Make a decision today to be more open minded and to look for the opportunities in every situation. Believe me, you will find them, and they will have a huge impact on your career. Whether you realize it or not, what you choose to see in every situation is totally up to you. Start asking yourself the two questions from this section and choose to see and seize the opportunity. These questions can be career and life changing, so please try to make them a habitual part of your thinking routine.

Two Questions to Maintain an Opportunistic Mind-Set

1. Where is the opportunity in this situation?
2. How can I learn and grow from this situation?

9.4 Understanding Your Role

Another important aspect of being a good leader is understanding your role. As a leader, your role is to help those that you are leading to do their jobs to the best of their ability. Therefore, in order for you to succeed, you should strive to help others succeed. This is a difficult concept for many engineering professionals to grasp. Many professionals are so worried about advancing their own careers that they shy away from helping others because they believe that it will take away from their own success. This is another limiting belief that can be detrimental to your career.

> As a leader, your role is to help those that you are leading to do their jobs to the best of their ability.

I have said over and over again in this book that in order to advance your career, it's important to always put yourself in a position to succeed. Helping others to do so is absolutely the quickest way to accomplish this. In our competitive mind-set, so many engineering professionals feel that they have to constantly fight for themselves and the idea of helping other engineers, whether they are coworkers or not, is perceived as a bad career move. Please don't take this approach. It will only bring you negativity and impede your success, and it may also cause many people to dislike you, thus making your career more miserable than enjoyable.

I have seen both types of professionals in my engineering career—those who focus solely on their own advancement and those who are dedicated to helping others. Guess what? The helpers always seem to have more rewarding and enjoyable careers. The helpers are those people that everyone in the office absolutely loves to work with, and they are the people that everyone wants to be around at social gatherings. The helpers are those professionals that have extraordinary careers and truly enjoy their lives.

Please don't confuse understanding your role with delegating, which I will discuss in the next section. Delegating is about knowing what project tasks you "should" or "should not" be doing and assigning the "should not" ones to others who are capable of completing them.

> Understanding your leadership role is knowing that your responsibility is to provide support to those that you are leading and then let them run with it.

I distinctly remember that as a young design engineer, I once had two managers—they were the exact opposite of each other. From my perspective, one of them understood his role as a mentor and guide and the other didn't. During a particular project, when I was finished with a design, one of the managers would sit with me immediately or soon after I finished and review it, while the other manager never had the time to review it with me. I thoroughly liked working with the one that understood his role, did the review immediately, and gave me the feedback, information, and confidence that I needed to finish the design. The other manager's apparent lack of interest always left me questioning how much he really cared about the project or the team, since he was always too busy to work with me. My productivity, efficiency, and level of enjoyment in the workplace were drastically higher working for the leader who took the time to go over my work.

Take a look at your own career. Are you supporting those that you are leading, whether they are staff members, project team members, or even classmates? Are you putting them in a position to succeed by giving them the materials, information, attention, support, and knowledge that they need to succeed, or are you solely focusing on your own tasks to ensure your own rise up the corporate ladder? Please think about the role you are playing in your career and the impact it has on yourself and others.

> "Leadership is about capacity - the capacity of leaders to listen and observe, to use their expertise as a starting point to encourage dialogue between all levels of decision-making, to establish processes and transparency in decision-making, to articulate their own values and visions clearly but not impose them. Leadership is about setting and not just reacting to agendas, identifying problems, and initiating change that makes for substantial improvement rather than managing change."
> —Dr. Ann Marie E. McSwain   [6]

9.5 Delegate, Delegate, and Then Delegate Some More

"Delegating" means to me to give or commit (duties, powers, etc.) to another as agent or representative. A synonym for the word delegate is "assign." (I find that understanding the meaning of a word or concept is important to technical people; we often like to start with a solid foundation and work up from there.) If there is one aspect of leadership that acts as a foundation for all of your leadership skills, it is the ability to delegate. Learning how to do this properly is one of the biggest challenges that engineers face in their careers, but when mastered, it will allow professionals to move from being good leaders to shining as a great leaders.

We, as engineers, are hands-on, analytical problem solvers. When we are faced with a challenge, problem, or task, we usually try to solve it all on our own. From our start in engineering, we were thrust into a world of numbers and complex equations. We know how to design elements such as gears, circuits, data sheets, and bridges, and we take pride in doing so.

However, a time will come in your career when you will approach a fork in the road, and you will be presented with the choice of continuing to do calculations and maintain technical responsibilities or moving on to a new set of challenges in a management role.

> Someday, there will be a fork in the road. You will have to choose to stay in your cubicle crunching numbers or walking down the hall toward management.

The minute that you take the path toward management, it is imperative that you learn how to delegate. If you move into a management position and don't take the time to learn how to delegate, you might as well go back to being a designer because you will be in for a very rough ride. You won't survive for long; but if you do, you may be a manager by title but lacking true leadership abilities. There is a big difference between simply getting the management job done versus being a great leader who has people clamoring to work on your projects.

To that end, here are some steps you can take to become an effective delegator.

First and foremost, you have to learn to "let go." If you are going to delegate effectively, you cannot try to control every aspect of every project, and the earlier on in the process that you accept that fact, the better. As a manager, you are now responsible for overseeing multiple projects rather than handling one project. Think about the benefits of being able to assign tasks to others that you previously handled and how many new opportunities this may open up to you. This is an important step, because if you can't embrace it, you will never feel comfortable giving the work to others. One book that has helped me to delegate more effectively is *The One Minute Manager* by Blanchard and Johnson [7]. The book provides strategies for setting clear expectations and getting high-quality results from your team.

When preparing to delegate, be sure to break up the project tasks into small achievable benchmarks that can be easily explained. Once you have broken up the tasks, I recommend assigning them to people that match up well with them. For example, if a specific computer program is needed to complete a certain task, assign it to someone who is comfortable using that program. Odds are that you will not always be able to choose the person for each task you need to assign. It is important that you take the time to explain each task in detail when assigning it to someone. Put your expectations in writing first through an e-mail, and then personally go through it together step-by-step to be sure that everyone is clear about responsibilities.

As well, it is also helpful to determine together how long it will take to complete the task, avoiding surprises in the schedule. Setting clear objectives also removes any element of surprise on your staff's behalf should you come looking for the completed task on the same day that you assigned it. Then, based on the expected length of the task, check in regularly and make sure that your staff knows that they can ask you questions as needed to, especially when they are learning something new. Even though Action Exercise Worksheet 11-4 in Chapter 11 of the book is laid out for using the SMART process for pursuing your goals, you can use the same process and worksheet to break down your projects into smaller, more manageable steps.

There will be times when you will have to delegate work to employees that are new in the field and have no prior experience. This can be extremely challenging and requires plenty of time and patience. Once you get good at it, it's like riding a bicycle—you can do it over and over again expecting good results. When delegating in this situation, you will most likely need to spend a fair amount of time at the start teaching and answering questions. You must make it very clear to subordinates that you know how it is to be in a learning phase, and not only is it okay for them to ask you questions, but you expect questions until they are comfortable with a certain task. Make it clear, repeatedly, that your door is always open and that you expect to see them in your office quite a bit for the first few weeks. Nobody wants to be labeled as a pest. By taking this approach, you will put those coworkers at ease and ensure information is shared, so that in the future they can handle the task without you. Do not make the mistake of trying to cut corners by doing some of the work yourself in order to save time. You are better off totally committing to supporting these colleagues until they understand; later, they can work without such close supervision.

The last step in delegating is to show appreciation once a task is complete. Be sure to thank team members for the work done and give compliments about aspects of the work that you especially liked. Now that the work has been successfully delegated and trained for, ask them if there is anything in the process that could have been done differently. You'll be surprised to learn that when people are doing something for the first time, they will often bring some of their own creativity to the table and help you optimize the process for future use.

Last, if individuals didn't perform the way you expected, have a conversation about that concern, but do so in a way that identifies specifically what was done wrong, how it differed from what was expected and agreed upon, and *how you will help them to improve going forward.*

I know that the ability to delegate is a difficult skill for engineering professionals to embrace, but I hope the steps provided in this section make it easier for you. Once you master this skill, you will look back and say, "Why didn't I do this sooner?" The time you will free up by delegating will allow you to seize and capitalize on new opportunities for both your company and your career.

9.6   Earn the Trust and Respect of Your Team

In order to be a powerful leader, you will need the support of your team. To gain support, you will need to garner their trust and respect. Being liked and being respected are two totally different things, and great leaders are usually both liked and respected.

The easiest way to earn the respect of your team members and staff is to respect them. People think that to get respect they have to like people or compliment them on the work they are doing. That's not true, and usually, people see through compliments that are given for the wrong reason. If you want to earn the respect of your team members, challenge them. Ask them about their career goals and dreams, then give them the opportunity to achieve them, even if it means letting them do something that they have never done before. Trust them. When my supervisors used to challenge me, I loved it, and I appreciated them for trusting in me. It really gave me a boost in my own confidence level, and I appreciated and respected them for giving me a shot.

Another way to earn respect from your team is to lead by example. If you believe in something, practice what you preach. Don't ask your employees to improve their personal dress code and then show up in jeans the next day. Don't teach them to prepare for a meeting a certain way and then arrive without being organized. This happens all too often in the corporate world, and it is one of the primary reasons that managers don't gain the respect of their staff.

Be open to new ways of doing things. Let's face it—as you become more experienced in your career, there may be younger professionals who have new ideas, new technology, or new methods for doing things. Be open-minded and even ask them their opinions about how to do things more efficiently. Not only might this feedback help you, but also they will respect you for asking their opinion and it will help to boost their confidence tremendously.

Take some time to think about the managers that you have worked for who you really enjoyed dealing with. You probably respected them a great deal. By recalling how they earned your respect, you may find strategies that you can implement yourself. It is always helpful to learn from the past or even to put yourself in the shoes of those you are trying to lead. Doing so will help you to gain a better understanding of what your team is dealing with so that you can guide the team through it in a positive and respectful way.

Apply some of these techniques throughout your career and make it a point to try to earn the respect of your team members. A big part of creating an extraordinary career is being respected by those that you work with, as that respect will be key in your leap from manager to leader.

> Being liked and being respected are two totally different things, and great leaders are usually both liked and respected.

9.7   There Is No "I" in Team

I mentioned earlier that many managers tend to think more about themselves than their teams, which is a major downfall that can impede them from becoming strong leaders. I recommend adopting a "team" mind-set from the beginning of your role as a manager. As a manager, you are absolutely a part of the team, but you can't view yourself as being any bigger or better than any other member of the team. You are all in it together. If one team member fails to meet a deadline, the entire team fails to meet the deadline, including you. Try to use words like "we" and "us" when you address the team rather than "you" and "me." I know this doesn't sound like an important leadership technique, but it is.

As much as you are a part of the whole, when something goes wrong, you should claim full responsibility. Yes, it is a team, and yes, you are part of it, but you are also the leader, and if the team doesn't perform, you are responsible. You can certainly discuss the performance of the team and challenge the members to improve, but the minute you blame them for a project that didn't work out, you will lose the respect that you worked so hard to gain from your team. If you are not comfortable taking the blame, you may not be suited to be a leader.

Just as you assume all of the responsibility when something goes wrong, you should also give all of the praise to your team members when something goes right. At this point, you may be thinking, "This leader thing is not all it's cracked up to be." Look at it this way; if your team performs well, you will be rewarded with praise and maybe even with financial remuneration. The next time that you see a high-profile powerful leader being interviewed after a victory or even a loss, such as a politician or an athlete, the responses will tell you a tremendous amount about that person's leadership ability based on whether or not she/he recognized the contribution made by the whole team.

> "It looks like a solitary sport, but it takes a team." —Diana Nyad, World Record Long-Distance Swimmer   [8]

Derek Jeter, once a young kid with a dream to play professional baseball, ended up becoming the captain of the New York Yankees and one of the all-time hit leaders in Major League Baseball. These achievements were largely attributed to Jeter's leadership abilities. I used to love watching him get interviewed after getting the game-winning hit. When the reporter asked him to describe the hit or how he accomplished it, he would always say something like, "This game wasn't about my hit. It was about our pitchers and fielders keeping the game close to put me in a position to have a chance to do what I did." He never spoke about his success; he only gave credit to others and spoke about his team as one.

You are now equipped with valuable information that you may put into action to develop your leadership skills. Manager or leader? It's up to you to decide which path you want to take.

9.8 Key Points to Remember

1. You ARE a leader. It is up to you to determine how well you lead.
2. Adopt a positive outlook in everything you do, say, and think. You can do this by surrounding yourself with positive people and by reading positive quotes, books, and articles. Also remember to ask yourself the two questions that will help you to maintain an opportunistic outlook at all times: "Where is the opportunity in this situation?" and "How can I learn and grow from this situation?"
3. Understand your role, which is to support those that you are leading by striving to always give them the materials, information, attention, support, and knowledge they need to succeed.
4. To delegate effectively, do not try to control every aspect of every project. Instead, assign tasks to others so that you can oversee multiple projects rather than handling every aspect of one project. Doing so will free up your time to create and embrace new opportunities.
5. When delegating, be sure to break the assignments up into small tasks, explain your expectations up front, and show appreciation when the task is complete.

6. Earn the trust of your team members by challenging them, leading by example, and asking them their opinion on your leadership habits.

7. Understand that as the leader, you are a part of the team and must act like it. Remember there is no "I" in team.

8. It is your responsibility as the leader to take the blame when things go wrong and to give praise to your team members when success is achieved.

A Boost from Your Professional Partner

Are you going off and trying to do too much again? To be a leader, you have to learn how to delegate. Wanting to do EVERYTHING will eventually catch up with you in more ways than one. Delegating is a great way to show others that you trust them and that you are there to help them to take their next step.
Start letting go of some of the tasks you know you shouldn't be doing.

Your Delegation Coach,
—Anthony

References

[1] J. Lowe, Jack Welch Speaks: Wit and Wisdom from the World's Greatest Business Leader. Hoboken, NJ: John Wiley & Sons, 2007.

[2] B. Schneider, Energy Leadership. Hoboken, NJ: John Wiley & Sons, 2008.

[3] E. J. Giltay, et al., "Dispositional optimism and the risk of cardiovascular death: The Zutphen elderly study," Archives of Internal Medicine, vol.166, no. 4, pp. 431–436, February 2006.

[4] P. Carrington, et al., "The use of meditation: Relaxation techniques for the management of stress in a working population," Journal of Occupational Medicine: Official Publication of the Industrial Medical Association, vol. 22, no. 4, pp. 221–231.

[5] H. Benson and M. Z. Klipper, The Relaxation Response. Reprinted with Permission for New York, NY: HarperCollins, 1975.

[6] CosmoLearning, Leadership [Online]. Available at http://www.cosmolearning.com/topics/leadership-418/ (accessed on April 30, 2014).

[7] K. Blanchard and S. Johnson, The One Minute Manager. New York: Berkley Publishing, 1981.

[8] "Diana Nyad Biography," The Biography.com website. Available at http://www.biography.com/people/diana-nyad-21329683 (accessed on April 14, 2014).

10

The Time Is Now: Take Action

The path to success is to take massive, determined action.

—Anthony Robbins [1]

10.1 The Time Is Now

So far, I hope you have gained a wealth of information that you can take and immediately implement into your engineering career and life. However, I want to warn you about something that you may encounter while doing so. You will face doubts along the way, whether it be self-doubt or others who doubt you or your abilities. Please don't let these doubts stop you. People may tell you that you are too young or too old to do things at this stage in your career. Disregard their "advice" and follow your own best judgment.

You have the ability to do anything you want in your career at any time as long as you put your mind to it and follow through. If that weren't the case, I wouldn't have taken the steps that I have shared with you in this book in my own career. I probably would not have even written this book if I did not feel compelled to take action and do it right away. If any kind of doubt that you have about creating an amazing career starts to creep into your mind, just think about all of the success you have had in the past and use those positive thoughts to propel you through the doubt.

Engineer Your Own Success: 7 Key Elements to Creating an Extraordinary Engineering Career,
First Edition. Anthony Fasano.
© 2015 The Institute of Electrical and Electronics Engineers, Inc. Published 2015 by John Wiley & Sons, Inc.

> You have the ability to do anything you want in your career, at any time, as long as you put your mind to it and follow through.

10.2   Do Not Settle for Less

Hopefully, I have helped you to set clear goals, define success in your own terms, and find practical ways to achieve those goals. As you move forward, it is important to always be true to yourself and not settle for less than your vision of success. Being true to yourself sounds courageous and honorable, and it can vary from person to person. To me, being true to yourself means living your life and developing your career in a way that allows you to pursue your dreams while staying aligned with your values and your beliefs.

I believe that too many people settle for less than they deserve in their career and life. Ask yourself these questions: Do you enjoy going to work each day? Are you challenged, engaged, and/or inspired in your career? If you are not, then I would ask you to honestly examine why your circumstances are less than ideal.

Unfortunately, western culture has forced many of us (thankfully not all of us) to adopt the belief that work isn't supposed to be enjoyable—it's all about getting a paycheck. I believe that all of life, both personal and professional, is meant to be interesting, exciting, and extraordinary. That doesn't mean that we don't encounter challenges along the way, but most of the time, these challenges present new opportunities for learning and growth.

> "This isn't a practice life."—Jeff Bluestein

During my graduation ceremony where I received my master's degree in civil engineering from Columbia University, the keynote speaker was the CEO of Harley-Davidson at the time, Jeff Bluestein. He rode into the ceremony on a beautiful chrome Harley-Davidson motorcycle. He stood up at the podium, in jeans and a ponytail, and started his address with the quote, "This isn't a practice life." He then spent the next 15 minutes delivering an inspirational speech on how we only get to go around this merry-go-round once, and it is our choice as to how we embrace it. That speech and quote had a profound effect on me, and I hope to spread some of that inspiration to you through the words I have written in this book.

I challenge you to take a good look at your career and ask yourself, "Is this really where I want to be?" If the answer is *No*, then set new goals today and start going from where you are now to where you want to be.

10.3   You Must Make Time for Your Own Development

I have spoken to hundreds of engineers that have either read my book or are members of my Institute for Engineering Career Development, and many of them tell me the same thing: "I don't have time to improve myself. I am too busy in my job."

If you constantly put your job and other responsibilities above your own development, well—to put it simply—you will never develop. You must hold your own personal development paramount to everything else in your career and life. There is a great quote from the book *The Monk Who Sold His Ferrari* by Robin Sharma [2] that describes my thoughts here perfectly:

> My friend, saying that you don't have time to improve your thoughts and your life is like saying you don't have time to stop for gas because you are too busy driving. Eventually it will catch up with you. [2]

So many people hold all of their work and personal deadlines in high regard, but they dismiss the importance of their own development. Please don't do this. The discussions in this book and the tools in Chapter 11 will facilitate action. It's time to get the process rolling!

10.4   Think Like an Entrepreneur in Your Career

As I near the end of this book, I want to give you one piece of advice that was one of the most valuable I received as an engineer; it influenced my decision to title this book *Engineer Your Own Success*. While it is imperative that you provide your clients with stellar service and give your current employer the best you have every day, it is also important to remember that you must always stay focused on your own goals. A client can leave you for another engineer at any time. In the blink of an eye, your employer could terminate you or go bankrupt. Engineers forget this all the time, and they work their careers away, never stopping to think about what they want and the skills they must develop to attain it.

> You have brains in your head.
> You have feet in your shoes
> You can steer yourself
> any direction you choose.
> You're on your own. And you know what you know.
> And YOU are the guy who'll decide where to go.
> —Dr. Seuss, Oh The Places You'll Go [3]

You should take an entrepreneurial approach in your career. Treat yourself like a business, strive to constantly grow, and get to the point where people need you. Think of skills you can develop that will make you indispensable. Seth Godin spends his entire book, *Linchpin*, discussing how you can become totally indispensable in a technology-driven world where people think indispensability is a thing of the past [4]. It's not. So whether or not you are a business owner, you are now. Build your business in a way that customers are lining up to buy you. Engineer your own success.

So whether or not you are a business owner, you are now. Build your business in a way that customers are lining up to buy *you*.

10.5 Take Action

Many professionals come up with great ideas that never become more than just that—ideas. They take classes upon classes or read books upon books, yet they never actually use the information that they obtained to do anything about taking their ideas to the next level. I don't want that to happen to you. When I find a book that I believe will help me reach my goals, I study it and actually designate a full notebook to notes and sketches on the book. These are my own shorthand notes that I refer to time and again.

Chapter 11 of this book will push you to create action items and clearly defined steps to achieve them. Some of these action items might include setting career goals, defining success in your own terms, signing up for an exam, finding a mentor, improving your public speaking skills, creating a list of 10 people that you want to start building relationships with, developing an effective to-do list, or joining a new professional society.

You can no longer say that you don't have the information you need to succeed in your engineering career. It is now up to you. I wish you the very best in your journey. For support, I would encourage you to sign up for the free newsletter that is both inspirational and informational, which I send to thousands of engineers every week. You can read about it and sign up for it at www.EngineeringCareerCoach.com. The newsletter will serve as a weekly reminder to stay focused in your career advancement efforts. Finally, I have included a Recommended Reading section in Appendix A, providing you with a list of some of the books that have helped me tremendously in advancing my own career!

Remember that it's up to YOU to take YOUR career to wherever YOU want to take it!

10.6 Key Points to Remember

1. Always remember that you have the ability to do anything you want in your career at any time as long as you put your mind to it and follow through.
2. All of life, on both a personal and a professional level, is meant to be interesting, exciting, and extraordinary. Don't settle for less than your ideal career and life. In the words of Jeff Bluestein, "This isn't a practice life."
3. Reading this book alone will not advance your career. It's now up to you to take action by using the tools and templates in Chapter 11 of the book to create action items to help you achieve your goals.
4. Run your career like a business and become indispensable.
5. Sign up for your free weekly inspirational newsletter at www.EngineeringCareer Coach.com to help keep you motivated and on the path to success!

> ### *A Boost from Your Professional Partner*
>
> *Clear goals, a good plan of attack, and the ability to surround yourself with positive people will set you up for success. Action will convert success from a goal to reality.*
> *You have planned enough.*
> *It's go time!*
> *No looking back!*
>
> Your Professional Partner,
—Anthony

References

[1] P. Bowden, Telling It Like It Is. CA: CreateSpace, 2011, p. 465.

[2] R. Sharma, The Monk Who Sold His Ferrari. San Francisco, MA: HarperCollins, 1996.

[3] Dr. Seuss, Oh, The Places You'll Go. New York: Random House, 1990.

[4] S. Godin, Linchpin: Are You Indispensable? New York: Portfolio Trade, 2010.

11

Tools and Templates for Setting and Achieving Your Career Goals

Engineer Your Own Success: 7 Key Elements to Creating an Extraordinary Engineering Career,
First Edition. Anthony Fasano.
© 2015 The Institute of Electrical and Electronics Engineers, Inc. Published 2015 by John Wiley & Sons, Inc.

11.1 Template for a Winning Résumé

Jane Doe, PE
123 Smith Drive
Smith, NY 90210
Phone: 510-555-5555 E-mail: janedoe@email.com

Summary/Objective
Include a unique summary that describes what you can contribute to the company. Also, use this section to include the job number or specific name of the job if applicable. If you have nothing unique to mention or specific like a job reference number, consider eliminating this portion of the résumé.

Professional Experience

Most Recent Company, City, and State (e.g., TX)   Start Month Year–Present
Position(s)
• List all your job responsibilities in bullet points, starting each with a verb.
Company 2, City, and State   Start Month Year–End Month Year
Position/s
• List all your job responsibilities in bullet points, starting each with a verb.

Professional Licenses, Memberships, and Affiliations
List all professional licenses, memberships, and affiliations. Spell out all abbreviations of the organization.

Society Activities
List your position, organization, and responsibilities. Include year term, as applicable.

Education (Note: This Section Goes at the Top for Students or Recent Grads)
Most Recent Educational Institution   Start Month Year–End Month Year
Location of educational institution
    Relevant coursework

Research
Provide information related to research, as applicable.

Awards and Certifications
List all the awards and certifications received starting with the latest and follow the following format:
Received (latest award or certification) from (award giving body) in (month year).

Computer/Internet Skills
List all the software and other skills in this section.

Activities/Hobbies
List hobbies and other activities but only if they directly impact/inform the job.

References
Aside from offering traditional references, remember that you can include recommendations in your LinkedIn profile. For example:
Over 20 recommendations on LinkedIn profile: http://www.linkedin.com/in/anthonyjfasano

11.2 Action Exercise Worksheet—Define Your Values

Define your values here by writing, in full sentences and complete thoughts, the core values that you hold. What principles guide you daily?

11.3A Action Exercise Worksheet—Define Your End Results in One Year

Ideally...
Where do I want to be in my career in one year? What would my day look like?
Why do I want to be there? What are the specific reasons for those goals?
What additional skills or training will I need to get there?
How can I ensure that I will achieve that goal?
Who can help me reach my goals?
How do my goals reflect my values?

11.3B Action Exercise Worksheet—Define Your End Results in Two Years

Ideally...
Where do I want to be in my career in two years? What would my day look like?
Why do I want to be there? What are the drivers for that vision?
What additional skills or training will I need to get there?
How can I ensure that I will achieve that goal?
Who can help me reach my goals?
How do my goals reflect my values?

11.3C   Action Exercise Worksheet—Define Your End Results in Five Years

Ideally...
Where do I want to be in my career in five years?
Why do I want to be there?
What additional skills or training will I need to get there?
How can I ensure that I will achieve that goal?
Who can help me reach my goals?
How do my goals reflect my values?

11.4 Action Exercise Worksheet—Formulate and Prioritize Goals

Goal	Timeline
1.	
2.	
3.	
4.	
5.	
6.	

11.5 Action Exercise-Worksheet—SMART Process to Achieve Goal #1

Goal #1:
S = Specific
List the steps required to achieve this goal.
M = Measurable
List at least three criteria that you can use to measure your progress toward your goal.
A = Achievable
While your goals should be challenging, ensure that they are realistic.
R = Relevant
Ensure that the steps you are taking toward your goal are relevant.

Contact at least one person who has achieved this goal or one very similar to it to make sure that your steps are both realistic and relevant to the goal.
Name: E-mail/phone no.
Name: E-mail/phone no.
Name: E-mail/phone no.
T = Time bound
Next to each of the steps listed at the beginning of this worksheet, write a deadline as a specific date. Add this specific date to your calendar with a few reminders.

11.5 Action Exercise Worksheet—SMART Process to Achieve Goal #2

Goal #2:
S = Specific
List the steps required to achieve this goal.
M = Measurable
List at least three criteria that you can use to measure your progress toward your goal.
A = Achievable
While your goals should be challenging, ensure that they are realistic.
R = Relevant
Ensure that the steps you are taking toward your goal are relevant.

Contact at least one person who has achieved this goal or one very similar to it to make sure that your steps are both realistic and relevant to the goal.	
Name:	E-mail/phone no.
Name:	E-mail/phone no.
Name:	E-mail/phone no.
T = Time bound	
Next to each of the steps listed at the beginning of this worksheet, write a deadline as a specific date. Add this specific date to your calendar with a few reminders.	

11.5 Action Exercise Worksheet—SMART Process to Achieve Goal #3

Goal #3:
S = Specific
List the steps required to achieve this goal.
M = Measurable
List at least three criteria that you can use to measure your progress toward your goal.
A = Achievable
While your goals should be challenging, ensure that they are realistic.
R = Relevant
Ensure that the steps you are taking toward your goal are relevant.

Contact at least one person who has achieved this goal or one very similar to it to make sure that your steps are both realistic and relevant to the goal.	
Name:	E-mail/phone no.
Name:	E-mail/phone no.
Name:	E-mail/phone no.

T = Time bound	
Next to each of the steps listed at the beginning of this worksheet, write a deadline as a specific date. Add this specific date to your calendar with a few reminders.	

11.6 Action Exercise Worksheet

Identifying the Right Mentor	
Type of engineer:	*i.e., civil engineer*
Discipline:	*i.e., structural engineer*
Job (be specific as possible):	*i.e., bridge engineer*

Review your goals on the previous pages; now, try to find an engineer with a job similar to yours who has achieved similar goals. Some places you might look include your company, LinkedIn, professional associations, and alumni groups.

Below is the text for an e-mail template that you can use to reach out to a prospective mentor (feel free to change it to sound more like your own voice). I would recommend that you do not mention the word "mentor" in the e-mail; your objective should be simply to try to set up an initial phone call. Many engineers think mentoring is much more work than it really is, so mentioning it in an e-mail may discourage a response. You can discuss mentoring on the call if it goes well.

E-mail Template

Hello FULL NAME,

I was referred to you by NAME (or found you through SOURCE) and wanted to talk to you about giving me some engineering career advice.

It looks like you have been very successful in the same engineering discipline that I am looking to thrive in. I was hoping I could ask you some questions about how to advance my own career.

Is there a phone number and a good time that I can call you in the next week?

Sincerely,

Joe Smith

Cell (xxx) XXX-XXXX

11.7 Action Exercise Worksheet

Mentoring Relationship Checklist for Protégés
❑ Establish your expectations in your first session with your mentor (i.e., you might say: "I am looking for you to help me pass my PE exam in the next 12 months").
❑ Meet with your mentor on a consistent basis, once a month minimum (preferably more frequent during the first few months).
❑ Spend the first few sessions with your mentor getting to know each other. You should also discuss your career goals as well as your immediate goals for the mentoring relationship.
❑ If your mentor does not bring up confidentiality, mention that you would like everything said to be in confidence. This should make it easier for you to open up on difficult issues.
❑ Toward the end of each mentoring session, commit to take certain steps before your next meeting.
❑ Ask open-ended questions during each session that can't be answered simply with "yes" or "no."
❑ As often as possible, create written documents with your goals and your action steps and share them with your mentor.
❑ Be courteous and respectful toward your mentor at all times.

11.8 Action Exercise Worksheet

Mentoring Relationship Checklist for Mentors
❑ Provide guidelines for the relationship early on including the time and location of meetings.
❑ Meet with your protégé on a consistent basis, once a month minimum (preferably more frequent during the first few months).
❑ Explain to your protégé that the relationship is confidential.
❑ Discuss how you plan on introducing accountability into the relationship (i.e., asking the protégé to commit to goals for each session).
❑ Ask your protégé to review personal career goals during the first session or help to set some. Offer advice to create some short-term goals that will be the focus of your work together for the next 6–12 months.
❑ Meet with your protégé in person from time to time if possible.
❑ Strive to be a resource for your protégé at all times.
❑ At the end of each session, ask your protégé to commit to some small goals that he or she can achieve before the next session.

12

Engineering Your Own Success Stories from Practicing Engineers

The following are stories that practicing engineers submitted to me after reading the first edition of this book. I wanted to share some of these stories so you can see how engineers are using the information to succeed. There is nothing more motivational than seeing your colleagues achieve their goals.

12.1 Planning to Be an Extraordinary Engineer

This first story is from **Eric Juma**, *a 15-year-old Connecticut high school student with an intense passion for science and engineering. He has wanted to be an engineer ever since he can remember, and now, in addition to his high school studies, he blogs frequently on the topics of engineering, nanotechnology, and medicine.*

Since I was very young, I knew I wanted to be an engineer. When I entered high school, I saw many opportunities to learn about engineering, but I could not figure out which ones to take advantage of and what steps I could take on my own to become an engineer. There is not much advice about pursuing engineering in high school outside of classes, so in my second year in high school, I was lucky enough to come across *Engineer Your Own Success*.

Engineer Your Own Success: 7 Key Elements to Creating an Extraordinary Engineering Career,
First Edition. Anthony Fasano.
© 2015 The Institute of Electrical and Electronics Engineers, Inc. Published 2015 by John Wiley & Sons, Inc.

The chapter that influenced me most was the first chapter about setting goals. From this chapter, I learned about the steps of setting goals for a career, which helped me make decisions about how to spend my time and prepare for studying engineering in college. This chapter isn't just useful for someone looking to study engineering—it provides career advice that can be useful to anyone from an elementary school student to a senior engineer. Thank you, Anthony, for guiding me toward what I now believe will be an extraordinary engineering career.

12.2 Realizing a Dream of Becoming a Structural Engineer

Ilias Gibigaye made a difficult decision to leave behind his career as a professional soccer player on the West African coast and journeyed to the United States to pursue his dream of becoming a structural engineer, like his father. He realized that dream in 2013 and now works for a civil/structural engineering firm in Oklahoma City where he has moved up the ranks rapidly and is involved in activities such as project design and management, business development, recruiting, training, and mentoring. Gibigaye has a Bachelor of Science in Civil Engineering (structural engineering concentration) from the University of Nebraska-Lincoln and is currently pursuing a master's degree in structural engineering at the University of Oklahoma. He is an articulate communicator and speaks French, English, Russian, and Ukrainian fluently. He has also been focused on refining his public speaking skills and leadership abilities by serving as president of his Toastmasters Club and as chair of the Oklahoma City ASCE Younger Members Group. These are his thoughts about engineering his own success.

The book *Engineer Your Own Success* has been instrumental in my professional career development. Through my father, who worked as a structural engineer, I had an early exposure to bridges, roads, and buildings and always dreamt of embarking on a career in structural engineering. Upon graduation from college, however, the economic downturn prevented me from getting into the structural realm, and I was working for an environmental engineering firm as an attempt to acquire some civil engineering experience.

Six months into my tenure, as my dream of becoming a structural engineer was gradually fading away, I came across *Engineer Your Own Success* author Anthony Fasano's engineering career development blog where I read a myriad of inspirational articles and subsequently purchased his book. After reading the book and then joining Anthony's Institute for Engineering Career Development (IECD), I felt rejuvenated and hopeful.

During our one-on-one coaching sessions where we discussed topics in the book, Anthony did an excellent job of ascertaining my goals, my life circumstances, my strong attributes, and areas that needed improvement and assisted me in mapping out steps to accomplish my objectives. He would keep me motivated and always instilled in me a positive attitude even when the chips were down.

From that point onward, the book became my career guide; Anthony became my mentor, my professional career coach, and, more importantly, my friend. He was always attentive to my preoccupations and extremely responsive. What differentiates Anthony from other coaches is not only his engineering background, but his ability to connect with people's humanity. Not only did he challenge me, but he also infused me with energy, innovation, and heroic performance. One of the best lessons I learned from Anthony and his book is the belief that I have the requisite potential to propel my career wherever I want.

The information in the book has also helped me to hone several professional skills such as time management, communication, public speaking, and networking. As a direct application of these skills, I have managed to develop a close relationship with a senior civil engineer who helped me clinch a structural engineering position with a company in Oklahoma City.

The *Engineer Your Own Success* book is one of the best investments I have ever made in my life, and meeting Anthony through the IECD was utterly rewarding at a personal level as well as in my professional career. Thanks to Anthony, I have managed to resteer the ship of my professional career. Now that I am on the right track, I can proceed forward with serenity and confidence every day, thanks to the skills I have built. I am ready to engineer my own success.

12.3   A Big Step Forward for an Aspiring World-Class Engineer

Mitch Guy, PE, whose story is below, is a licensed professional engineer in Shreveport, Louisiana, where he is striving to achieve his ultimate goal of becoming a world-class engineer. He has had experience in the field of transportation engineering with a focus in airport and railroad design but in 2013 was tasked with overseeing a massive wastewater project that includes the inspection over 100 wastewater pump stations. His responsibilities on this project include physically evaluating, reporting, and making recommendations for each of the stations. Guy has a Bachelor of Science in Civil Engineering from Louisiana Tech University.

I have always known I wanted to go far in my career; I just never really knew what it meant or how to do it. Every other engineer I have ever known just wanted to do their jobs from 8 to 5, go home, and get paid; so I've always felt kind of alone in wanting to build a magnificent career.

I met Anthony at a civil engineering conference in Nashville, TN, where he was doing a mini-session on networking. At the end of the 15 minute presentation, he made his book *Engineer Your Own Success (EYOS)* available to us. To be honest, I really just purchased a copy to have something to read on the plane ride home. Once I started reading, though, I discovered two things that really stood out to me: (i) Anthony was the first engineer I had ever heard of who valued his career the same way I did, and (ii) the book contained a chapter on credentials with an entire section dedicated to passing the Principles and Practice of Engineering (PE) exam—an exam I would be sitting for in just over 2 months.

When I started reading *EYOS* at the airport, I knew by the time I finished the introduction that this book was going to change my life. I had read self-help books before, but never one by an engineer for engineers. Engineers are a rare breed; it literally takes one to know one. By the time I was done with the book, I knew I wanted to be a *world-class* engineer—an engineer with an outstanding reputation as both a technical expert and a professional in his or her field. For example, I wanted to be an engineer who can come up with smart, innovative designs and/or motivate a team to develop similar designs on a grander scale; that would be my definition of world-class. Reading this book helped me identify this goal as well as to begin the process of making it a reality.

The fact that I passed my PE exam on the first attempt is testament that *EYOS* offered me some great advice for my exam preparation. Advising me to bring a pocket dictionary to the exam was one of the simplest, yet greatest, tips I would acquire. I remember one particular question on the exam that contained some unfamiliar terminology; once I looked the words up in the dictionary, however, I found the answer to the question was quite simple.

The book also encourages examinees to network with each other to help prepare for the exam. This inspired me to assemble a group of examinees from different engineering backgrounds to come to my office one night and discuss some of the exam material. We spent 3 hours just talking about some of the topics we each found worrisome. I had a lot of questions for the structural engineer who had her own questions about transportation engineering and so on and so forth. We were all able to help each other, and that short time after work turned out to be one of the most informative moments of my entire exam preparation.

I would recommend *Engineer Your Own Success* to any engineer wanting to make the most of his or her career. It is a great starting point for people to discover themselves as professionals, to determine specifically what they want from their careers, and to define the steps required to get there.

12.4 A Boost of Confidence to Spur Maximum Potential

*Here, **Diego Espitia** provides some inspirational words. He is a civil engineer in Miami, Florida, focusing in the area of transportation design. He achieved his high school goal of becoming a civil engineer and is now progressing through the transportation industry. Espitia has a bachelor of science degree in civil engineering from Florida International University.*

I knew I wanted to be a civil engineer since I graduated from high school. Therefore, as a sophomore in college, I decided to get an internship to get an advantage over everybody else. I figured, the earlier I know the business, the more opportunities I would have when I graduated. I sent applications to every firm I heard of. I always saw myself doing tall buildings. However, the company that gave me the opportunity as an intern was a roadway design firm. At that moment, I was just happy to be an intern somewhere. But with time, I increasingly enjoyed every roadway design job that I worked on.

Since then, my goal has been to be one of the best roadway design engineers.

There was a problem, though; a couple of years after graduation, I felt that I was not moving forward with my career, as if I was not being challenged enough. That is when I decided to go to another company that promised more responsibility and more challenges. To be honest, I sold myself very well in this firm. But I did not have the tools to be successful there. After a very tough six months in this firm, I went to Anthony's seminar at the University of Miami. After the seminar, I bought his book *Engineer Your Own Success* and started implementing all of his the steps and advice. I can honestly say that, thanks to his book, I gained confidence and I started to perform at my best.

I recently started working at another firm, and I am implementing all the knowledge from the book to help the firm grow. I am very thankful for everything that I learned from Anthony's first book and hope to use all the advice from the updated edition.

12.5   The Push Needed to Take Action

Pat Sweet is an electrical engineer working in the rail industry in Kingston, Ontario, Canada. There, he works as a lead engineer on new rail vehicle development projects. Sweet earned his bachelor of engineering degree from Dalhousie University and is currently pursuing his MBA from the Royal Military College of Canada.

Anthony's book, *Engineer Your Own Success*, has been an career goldmine. His advice has been instrumental in guiding me through my career in its early stages, helping me to prepare for continued success.

Engineer Your Own Success has guided me in some important ways. It helped me make the decision to finally take the leap and start working on my MBA after years of indecision. I'm not even halfway through my degree, and I've already seen big gains in my career. The book gave me the push I needed to take action, and I'm grateful that it did.

The book has also helped me to see the important leadership role that engineers have to play, regardless of what their job title is. Still being in the early throes of my career, this has been incredibly motivating for me. It has helped give me the confidence to step up and take action at work and demonstrate that I can lead, even as a young engineer. As a result, I've been promoted to be the lead engineer on a complex international project worth over $100 M, which is something that most engineers in my company don't take on without at least a decade more experience than I currently have.

Most importantly, though, Anthony himself has been an inspiring figure for me. His guidance, both in the book and in person, has been a great blessing. He encouraged me to take on big challenges and to grow myself as an engineer and as a person. I'd recommend *Engineer Your Own Success* to anyone.

12.6   I Decided to Start Planning for Me in My Career

Denise Nelson is civil environmental engineer (PE), LEED Accredited Professional (LEED AP), and Envision Sustainability Professional (ENV SP) in Richmond, Virginia. She has 11 years consulting experience in collection systems, water distribution, and sustainable infrastructure in the United States and overseas in New Zealand. She is active in many professional societies including serving in leadership roles for ISI, APWA, and WEF. Nelson received her bachelor of science degree in civil engineering from Virginia Tech and her Master of Science in Environmental Engineering from the University of Cincinnati.

The clear, straightforward guidance tailored for engineers in Anthony's book motivated me to consider other options in my career. I found the goals chapter (appropriately located at the beginning of the book) to be most useful. I, like most engineers, stood out in middle school for my math skills and was placed in a college prep track for targeting a degree in engineering. Throughout college and into my career, all of my professional development studies and activities started from the same point: my current position. I was involved in all the relevant professional societies, presenting at conferences, focusing on projects to expand my technical expertise, and developing my leadership skills. By all accounts, I was a highly successful engineer for my level. But after reading the goals chapter and joining an Institute for Engineering Career Development (IECD) webinar on goals, I decided to start planning from a new point: me.

I took some time to revisit my values, reflect on activities that I enjoyed doing, and envisioned the life I would like to have. I realized that my current path had been a great path for me, but after 10 years and life changes (marriage and kids), it was time to revisit and redefine my path.

Fresh goals gave me a new perspective when approaching the remaining chapters in the book. Again, I appreciated the straightforward guidance for entering new territory. This helped me develop a logical approach and a clear plan that minimized my anxieties about changing my career. Whenever I feel overwhelmed or doubtful, Anthony's daily and weekly e-mails also remind me to be confident and fearless. Plus, I have the support of the entire IECD for advice in every step of the process (including urgent situations!). Now, I am confidently on my way to a career that is a better "fit" for me, and I think that is the key to success!

13

The Best of the Blog

I maintain a blog at EngineeringCareerCoach.com where I have written hundreds of articles related to engineering career development. In this section, I share some of these articles that have been most helpful to engineers based on their feedback. Note that some wording here has been edited to take it out of the immediacy of the blog format.

13.1 What Is Your Ultimate Career Goal? (September 10, 2010)

Did you ever take the time to stop and think, "What is my ultimate career goal?" It's an open-ended question with tons of possible answers, but it will certainly force you to think about your career and where you are headed.

I am writing this post in response to a question that an engineer asked at one of my career advancement seminars. The question was something to the effect of, "How does my role fit into the big picture of the world?" What a great question! Unfortunately, the only person that can truly answer this question is the person that asked it; however, I thought I would share how I would answer this question.

Engineer Your Own Success: 7 Key Elements to Creating an Extraordinary Engineering Career, First Edition. Anthony Fasano.
© 2015 The Institute of Electrical and Electronics Engineers, Inc. Published 2015 by John Wiley & Sons, Inc.

I would start by brainstorming on the first question: "What is my ultimate career goal?" Do you want to make a lot of money, be president of a company, work as little as possible and enjoy life outside of work as much as you can, own your own business, and travel around the world during your career?

Next, I would follow up that question with three letters: *Why*? Why do you want to make a lot of money? Is it to support your family, or is it so that you no longer have to work? Why do you want to travel, or better yet, why wouldn't you want to travel in your career?

Once you figure out your goal and why you want to achieve it, then it may be much easier to determine your role or purpose in the world.

I hope this post helps you to think about your big picture and start taking action today to paint that picture in whichever way brings you the most success and happiness!

> *My ultimate career goal is to help engineers and other professionals in unleashing possibilities to create a career that is exciting, enjoyable, successful, and well balanced. I plan to do so in a way that I can still fully enjoy my life with my family as well as an exciting, enjoyable, successful, and well-balanced career.*

13.2   From Design Engineer to Manager in 2012: You Can Do It! (January 4, 2012)

I have received many questions recently through the different social media outlets to the effect of, **"How do I go from design engineer to manager in my engineering career?"**

It's a great question and one that many engineers ask. In my career travels as a design engineer, and the last few years as an engineering career coach, I have worked with, coached and spoken to many engineers about this topic specifically, and in this post, I want to offer some engineering career advice based on my experiences.

From Engineer to Manager

Learn How to Delegate

What do I mean by *learn* how to delegate? Well, I could have just said, start delegating, but many successful engineers will tell you that it's not that easy. As engineers, we get so wrapped up in day-to-day technical aspects of projects that when it's time to go take on a managerial role, we either don't want to give up the technical tasks to someone else or we are so involved in our projects that it is hard to take a more "hands-off" role.

Many engineers have the mentality of, "I want to do it to ensure it gets done correctly." That's understandable, as you are a competent project manager; however, to make the transition from engineer to manager, you are going to have to let other people help you.

You are going to have to take the time to teach them how to do it in order to free up your time for other things. The best way to do this is to start by giving small tasks to your team members to let them gain your trust. Once they build up your trust, you will feel more comfortable giving them larger tasks until you can remove yourself from the "trenches" and take on more of a managerial role.

Learn How to Talk to People in a Positive Way

I always tell engineers that the way you say something to people is more important than what you say. As an engineering manager, you will have to delegate to your team members and have regular conversations with them about what tasks you would like them to accomplish.

Many engineering managers think only about the success of their project and not the success of their people. When you delegate tasks to your team members, explain to them why you want them to do something, how it's going to help the project, and also how it is going to help their engineering career development. I have coached so many engineers that don't understand why they are not getting the most out of their staff, and when I review their e-mails to discuss their conversations, it's easy to see that their tone and choice of words is anything but inspiring.

No one has time to read a book these days, but to help improve your people skills, I recommend listening to the audio version of Dale Carnegie's best-selling book, *How to Win Friends and Influence People*.

Become a Great Presenter

One of the biggest misnomers among recent engineering graduates and younger engineers is that you don't have to be good at writing or presenting. That is 150% false! In fact, successful engineers are typically very good writers and/or presenters.

To make a successful transition from engineer to manager, you must be able to present your ideas clearly to a group of people whether it is a group of two or 200. You may have to present at a Town Board meeting in seeking project approvals, make a sales call to a prospective client, or speak in front of one of the local professional engineering (PE) societies. In all of these cases, your ability to present will have a profound impact on your success as an engineering manager.

The good news is that public speaking is not a talent that you are born with or without; it's a skill that can be learned. I urge you to develop this skill as early in your engineering career as possible. To improve your speaking skills, I recommend joining a local Toastmaster chapter and listening to the audio version of the book entitled *Speak to Win* by Brian Tracy.

So, in response to the question of how to become an engineering manager, I would say: learn how to delegate to others, learn how to talk to people in a positive way always thinking about their success, and, last but not least, become a great presenter!

13.3   Twelve Rules of Zen Monks That May Help You Reduce Stress and Improve Quality in Your Engineering Career (June 5, 2012)

Like many professions today, engineering can be a stressful profession due to the project deadlines, shrinking budgets, and client demands. In this post, I want to provide some principles that Zen monks follow and how you might apply them in your engineering career to help you reduce some of the day-to-day stresses and be more engaged and in the moment (or present).

I have to start this post by saying that it was inspired by a wonderful post that I read titled "12 Essential Rules to Live More Like a Zen Monk" by Leo Babauta. Leo Babauta is a simplicity blogger and author. He created Zen Habits, a Top 25 blog (according to *TIME* magazine) with 200,000 subscribers; mnmlist.com; and the best-selling books *Focus, The Power of Less*, and *Zen To Done*.

I want to give Babauta full credit, as I will be using the 12 rules from his post for the basis of this post; however, as you will read below, I attempt to take these 12 rules and describe how engineers can use them on a daily basis in their engineering career development:

1. **Do one thing at a time.** Zen monks do only one thing at a time. Babauta references a Zen proverb, "When walking walk, when eating eat." We as engineers multitask for a living! However, when we multitask, there is a high probability that the quality of our work may go down simply because we are not focusing 100% of our attention on it. This doesn't mean you can't have multiple active projects; it just means you literally do one thing at a time throughout the course of the day.

2. **Do it slowly and deliberately.** If you are able to give up multitasking and focus on one task at a time yet speed through each task, you will still not be giving it 100% of your attention. Slow down! As he says in his post, "Make your actions deliberate, not rushed and random. It takes practice, but it helps you focus on the task."

3. **Do it completely.** This may be one of the most critical rules of the 12 for engineers. How many times have you stopped in the middle of designing a gear or a beam or a stormwater basin to jump onto something else? You then come back to the design task at a later time and take at least an hour just to figure out where you left off. Do your best to finish off tasks completely before moving on to the next one.

4. **Do less.** This is an interesting one. Babauta discussed how Zen monks are not lazy, they are actually busy all day, but they only take on a few tasks so they can do them in accordance with rules 1, 2, and 3. I am not sure how well this would go over in the engineering world—doing less? What I can say though is we can be smarter about planning our day. Create realistic to-do items for the day so you are not trying to rush through a task in two hours, just to get to the next one, when in reality to produce a high-quality design, you need four hours.

5. **Put space between things.** Many engineers, in thinking about high efficiency, set their day up in a way where they leave no space between tasks; I admit that I do this myself. If I am coaching an engineer at 1 p.m. and have to schedule another coaching session that day, usually, I will shoot for 2 p.m. to "get all of my meetings done" at one time. However, I often notice that even though I know the 1 p.m. session won't go longer than 60 minutes, the 2 p.m. session is still in the back of my mind during the earlier session. I will be spacing out my sessions going forward, which I believe will make me more present with each client.

6. **Develop rituals.** Develop rituals throughout your day. This is different than number 7 because when you think about ritual, it refers to the way something is done. For example, if you have team meetings each day, develop a ritual or process for that meeting to make it a successful one and then follow that same procedure every meeting. Creating rituals throughout your day gives an item a certain level of importance and forces those involved to focus on following the procedure.

7. **Designate time for certain things.** This one should be easy for engineers as we are creatures of habit. Creating specific times for tasks throughout your day ensures that these tasks get completed on a daily basis. For example—and I know engineers will love this example—do your timesheet at the end of each day! Then when Monday morning comes, you won't have to spend two hours figuring out what you did last week.

8. **Devote time to sitting.** Having time each day dedicated to sitting or meditation is a staple of the Zen philosophy. There aren't many engineers that I know that meditate, but it is excellent practice for being present in the moment. If you struggle with meditation, do something in its place on a regular basis. For example, in his post, Babauta states, "I use running as a way to practice being in the moment. You could use any activity in the same way, as long as you do it regularly and practice being present."

9. **Smile and serve others.** Zen monks spend a part of their day in service to others as many engineers do, whether it's your clients, local community, or your kids! Serving others helps you to avoid being selfish in your daily actions. Smiling is an easy way to serve others and bring joy to someone else's day. He also cites volunteer work for charity as another way to serve others. For engineers interested in volunteering, I recommend a great organization called Engineers Without Borders.

10. **Make cleaning and cooking become meditation.** If you are a full-time engineer, you may not get the opportunity to do a lot of cooking or cleaning around the house; however, this is a wonderful way to practice being in the moment. You can use your time cooking and cleaning to focus 100% on the tasks at hand and ultimately help you to become more present.

11. **Think about what is necessary.** Zen monks don't have many things in life that they don't absolutely need. Now I am not asking you to become a Zen monk, because most likely you are not, but I am challenging you to take a look at

everything you own and think about what is really necessary. You might find it liberating to part ways with some of your material objects.

12. **Live simply.** Zen monks live extremely simple and embrace those things that are essential to them. For this point, I want to close with Leo's closing line because I believe he says it best, "There is no law saying what should be essential for you—but you should consider what is most important to your life, and make room for that by eliminating the other less essential things in your life."

If you want a certain thing, you must first be a certain person. Once you are that certain person, obtaining that certain thing will no longer be a concern of yours.
                                                                        —Zen proverb

I am on my way to hopefully become more Zen in my ways, and if you are interested in doing so yourself, I hope these 12 rules will help you to do so!

13.4   It's My Birthday! Who I Am Away from Work and Important Lessons That I Have Learned (August 26, 2012)

Today is my birthday! I am 34 years young and I wanted to take this opportunity to show you that I am not strictly just about helping engineers create extraordinary careers. I want to share with you some of the other things about me, mainly my life experiences to date and some of the more important lessons that I have learned along the way. I want to show you who Anthony really is.

Photo 1: I grew up in a small suburb of New York—the Village of Suffern in Rockland County. I played a lot of sports and my dad always coached me. Playing organized sports taught me a lot about teamwork. In looking back, I realize now that sports for kids should never be about winning and losing, **only about growing and learning.** Starting in a few weeks, I will be coaching my six-year-old daughter in soccer and will use this lesson wisely.

Photos 2 and 3: I come from a large Italian-American family. I have two younger brothers and we had a lot of fun growing up and still do when we get together. All we needed were three hockey sticks, a ball, and our driveway, and we were good for hours. We have 16 cousins on my father's side. We ate dinner every Sunday with all of them at my grandma's growing up. **I learned that family is as strong as you make it.** My parents made sure we saw my cousins weekly, and to this day, I have strong relationships with all of them. Just a note: today is Sunday and we will be bringing my kids to my parents' house to celebrate my birthday.

Photos 4 and 5: I went to Don Bosco Preparatory High School and then Lafayette College where I received a B.S. in Civil and Environmental Engineering. My parents taught me to **get the most out of every experience that you go through**, and I did—getting involved in many different clubs, sports, and activities in school.

Photos 6 and 7: I studied abroad for a semester in Brussels, Belgium. My wife Jill, who is a civil engineer too (girlfriend at the time), came as well. It was probably one of my fondest memories and most amazing experiences in life to date. We visited 13 countries during the time we were there. **Studying abroad opened my eyes to the fact that there were other countries and cultures out there besides my own.** I attribute much of my creative thinking, open-mindedness, and entrepreneurial spirit toward those six months abroad.

Photo 8: The best thing I did in college was meet my wife Jill. We have been married for almost 10 years now. When I told her three years ago (when she was pregnant with our second of three) that I wanted to leave a comfortable engineering job to go to executive coaching school and then help engineers create extraordinary careers, she looked at me with a blank stare for a minute but then said, "Do it." She has supported me ever since without blinking an eye. **I have learned that doing what you want to do in life at all costs and having your spouse fully support you is the best feeling in the world.**

Photos 9 and 10: We have three wonderful kids ages six and under. I never knew what to expect when becoming a dad, but now I know. Your kids just want you to listen to them and be there for them and that's what I try to do every day. I have learned that a high-paying job in a big city, a big house, and a luxury car mean absolutely nothing to me, and if I never have them, I'll be happier for it. **What means everything is the time you spend with the ones you love, the connections you make with them, and the experiences that you will remember forever.**

Photos 11 and 12: I enjoy going outdoors when I can. Photo 11 shows my wife and I at the Grand Canyon—what a sight! I also like fishing because it slows things down. Being out in nature can really ground you and reduce any stress that may have built up during your day or week. **I have learned that slowing life down allows you to better enjoy it one second at a time.**

Photo 13: The Anthony in this photo is the one that most of you know—the inspirational guy that travels the country urging engineers to create their own success by developing all of their nontechnical skills. I will still be that guy and won't stop until I have reached every engineer that I can find. In the meantime though, now you know a little bit more about me. **I have learned that getting to know people on a deeper level can be very rewarding.**

13.5 What to Do in Your Engineering Career When You Don't Know What to Do (May 30, 2013)

This post was inspired by a recent conversation that I had with one of our Institute for Engineering Career Development (IECD) members. She is a young engineer and not sure what direction to go in her engineering career. She is considering various questions, such as these:

- What specific discipline in her engineering field should she go into?
- What types of projects does she actually want to work on, on a daily basis?
- What kind of master's degree should she pursue?
- What will make her happy in her engineering career?

These are all important questions to be considered in your engineering career development, and to be honest, it's quite scary to have to deal with them all at such a young age. What happens if you answer one of them wrong? What if you go into a discipline and don't like it? What if you get the wrong master's degree?

My recommendation for anyone who is dealing with questions like these would be to try to build *flexibility* into your engineering career development plan. Try to take steps that help you to develop but still leave you open to other possibilities. Here are a few ideas of how to implement flexibility into your career:

1. **Selecting an employer.** If you are unsure about exactly what engineering discipline to pursue, attempt to get a job with a company that provides services in different engineering disciplines. By doing this, you create flexibility—if you don't like one discipline or department, you might be able to transfer to a different one within the same company.

2. **Consider company size.** The size of the company you work for can certainly dictate the types and variety of projects you get to work on. Smaller companies may give you an opportunity to be involved in many different facets of certain types of projects, while larger firms may expose you to many different kinds of projects and clients. When unclear about your future, select a company size that allows you to maintain flexibility as you develop.

3. **Consider the future opportunities a master's degree may yield.** Deciding which type of master's degree to obtain can be extremely difficult when you are unsure about your future; however, you might still consider pursuing a degree that provides more flexibility, such as an MBA or even a master's degree in engineering management. Of course, you must ensure that you obtain the degree you need for your desired credentials, but a business or management type of degree might be beneficial regardless of the engineering discipline you choose to pursue.

4. **Follow your passion as best you can.** Even though you are unclear about where you ultimately want to be, you most likely can still attempt to pursue jobs or disciplines that will allow you to do the things you enjoy. If you like working outside, find an engineering job that allows you to spend time outdoors. If you like working with people, explore disciplines that require a lot of interaction. This is important for maintaining a high level of enjoyment and engagement in your career.

5. **Be yourself. If at any time in your career you feel uncomfortable doing what you're doing, do something else.** Do something else regardless of how much work it will take or how many things you have to *Undo*. Nothing is worse than going through the motions in your career just because it's easier than changing.

This is a tricky topic, and it's hard to give black-and-white advice on something so unclear, but I hope some of the tips help you to build flexibility into your work and ultimately give you some direction in your engineering career.

13.6 Preparation Is Key to Engineering Balance in Your Career and Life (July 25, 2013)

One of the biggest challenges for many members of the IECD is time management, otherwise known as work–life balance. This is actually a challenge that most engineers face on a daily basis.

What Is Time Management?

Before I discuss improving your balance, what exactly is time management? Personally, I think it is a phrase that is defined differently by everyone who uses it and often is used as an excuse by people who aren't very efficient. Inefficient people can always just say, "I need to improve my time management skills." Who doesn't? You are always going to have more to do than the allotted amount of time in the day allows, so there will never be a time when your time management won't need improving. I recommend focusing on ways of better using the time you have.

Distractions Waste Your Time so There Is Less to Manage

I attended a conference recently, and on the two days I spent traveling to and from the conference, I accomplished more than I usually do in four days in my office. Why? Several reasons—the most important being that I was prepared. But also, there was no Internet on the airplane, no one called me, and no one came into my office for any reason. I guess in summary there are fewer distractions. The distractions that you encounter during the day limit the amount of time you have to be productive. Therefore, the fewer the distractions, the more time that you actually have available to work. We don't need to manage our time better; we need to spend more time in a focused state throughout our days.

It's All in the Preparation

Let's say we do have blocks of time in our day where we can focus, like me on my trip last week. We must ensure that we can get the most out of those times. In my case, before leaving for my trip, I had a list of things that I wanted to accomplish during the week. I made sure that some of the tasks were computer tasks but others were reading or writing tasks. For example, from the time you get on the airplane until you get up to a certain altitude (about 30 minutes), you can't turn on electronics. I spent that time reading and doing document review/markup. I performed the same tasks on the 30 minute taxi ride from the airport to my hotel. By simply taking some time at the beginning of your day or trip, you can ensure that you maximize your time. Failure to do so can cause an inordinate amount of stress while you sit on the airplane and think about everything you will have to do later on in the day or week.

How About Downtime?

My point with this post is not to say that you have to be productive during every second of every day; however, from my experience, when you maximize your time during the day, you increase your chances for a relaxed evening and night's sleep. I slept very well those two nights last week, because I knew I had accomplished a lot on those days. Downtime is important, but might not be relaxing unless you have accomplished a certain number of tasks that day.

The bottom line here is that everyone manages their time differently. I want this post to make you think about how you manage or spend yours. Are you staying focused during the day and enjoying relaxed evenings, or are you allowing yourself to get distracted during the day and then finding yourself very stressed at night, feeling that your time management is lacking?

Be prepared in the morning. Be productive during the day. Be at peace in the evening.

13.7 Six Ways to Reinvigorate Your Engineering Career Development (July 31, 2013)

We recently painted our kitchen cabinets a light-gray color. They are old wooden cabinets that we originally painted blue when we moved into the house about seven years ago. The change in color has brightened up the entire room and the dining area and has made the room much more enjoyable to be in. This got me thinking—how can we freshen up or reinvigorate our engineering career development efforts?

Do Your Own Performance/Career Review

As engineers, we are very hung up on the annual performance reviews our companies give us. Why not do your own review? Take a few minutes to review your recent career development progress. Also review your goals and how close you are to achieving them. Ask yourself if you are doing everything that you can to try to achieve your goals. If not, make adjustments. If for some reason you have not set clear career goals, then consider taking the time to do so. At the IECD, we are considering giving members the ability to fill out a yearly career plan to assist in tracking career development progress.

Seek Out Some Different Perspectives on Your Career and Life

We are very biased when it comes to ourselves, for obvious reasons. It is important in your engineering career development efforts, as well as in life in general, to seek out other perspectives. After you review your own performance, discuss your progress/goals/challenges with other people, whether they are career coaches, colleagues, or family members, or even through online forums. IECD members often pose questions about their specific goals and challenges on our private forums, and they have told us that the feedback has been invaluable in helping them to get the most out of their engineering career development efforts.

Reorganize Your Office

It's good to change things up once in a while. For some of you, changing things up might mean a new job; for others, it could simply mean reorganizing your office. Move some furniture around, paint the walls, bring in some plants, and declutter your space. All of these things can have a tremendous impact on your daily approach, attitude, stress levels, and overall development.

Read a Personal Development Book

It's amazing to me how reading a book can change your life and instantly inspire you to take massive action in your engineering career development. I can attribute my rapid engineering career development largely to a handful of powerful books. It's so easy to consume a book today, whether an audiobook on your commute, an e-book on your kindle, or a classic paper-and-ink volume in your spare time. Be sure to spend your time wisely on a book that will help you on your path to achieving your goals. In the IECD book club, we are currently reading *Think and Grow Rich* by Napoleon Hill, and I recently reread *The Alchemist* by Paulo Coelho. Both are classics, in my opinion.

Take a Vacation Outside of Your Country

I know. What does a vacation have to do with your engineering career development? A lot! By visiting another country and therefore another culture, you will open your mind up to new things, people, and places. I studied abroad in Brussels, Belgium, in 1998 as an engineering student. I was originally leery of leaving my college buddies for six months; however, the six-month, 13-country trip was one of the most amazing experiences—if not the single most amazing experience—of my life. I developed a new perspective on life, travel, and culture that helps me in my career and personal development efforts to this day.

Take a Risk: Big or Small

Your engineering career development can be accelerated tremendously by one scary word: RISK. Sometimes, in order to make a big move in your career, maybe even to go from engineer to manager, you will need to take a risk. Whether it is relocating within your company or to another company on the other side of the world, a risk can often greatly facilitate your progress toward your goals. I took one four years ago when I left my design engineering career behind, and if I hadn't, I wouldn't be writing this article and helping thousands of other engineers.

13.8 The Only Stability You Have in Your Engineering Career Is You (September 24, 2013)

If you are reading this post, you're most likely an engineer working for an engineering company. You get up each morning and go to work in what appears to be a "stable" job. In reality, though, no matter how safe or stable your job appears to be, your employer can decide at any time to lay you off and take away that perceived stability. However, your skill set, your professional network, and your reputation can allow you to react to a situation like this and quickly replace that paycheck. Developing your engineering skills as well as your nontechnical skills will give you the tools to truly maintain a stable engineering career. The following are some things you can do in your engineering career to ensure that it is genuinely stable.

Realize the Importance of Your Own Development

Don't get tricked by the illusion that your 9-to-5 job and paycheck create a stable career. Don't build your entire financial future around that paycheck. Go through your career with a plan B that addresses the question, "What will I do if I lose my job tomorrow?" Having a good answer to this question and a good plan to go along with it is where true stability will come from.

Build a Very Strong Professional Network

A company can always terminate your employment, but it cannot take away your relationships. If you build strong relationships in the industry, it is very likely that should you lose your stable job you will be able to quickly find another one.

I can't stress this point enough, as many engineers rank networking low on their list of things to do because they feel that they can lean on the relationships of their company. WRONG. If you take this approach and lose your job or decide to relocate, you will not be able to lean on the company's contacts. Also, having a strong network gives you a direct avenue to potential new business for your employer, which is like gold, especially in today's economy.

Become Known in Your Industry

You want stability in your engineering career? Become known in your industry. Speak at conferences on technical topics that you have expertise in. Join and participate in your local professional associations and community groups. This is how you build a strong reputation in your industry/community. Should you do this, if you lose your job or decide to look for another opportunity, employers will be lining up to talk to you. That is what I call stability.

Practice Top-Notch Customer Service

If you are at the point in your career where you get to interact with your clients, practice top-notch customer service at all times. Be friendly and kind to your clients and be superresponsive. Ensure that you are always honest and up front with them when telling them how long projects are going to take to accomplish and how much they are going to cost. Make helping your clients your highest priority. For example, if you see a project that may be of interest to them, forward them the information. If you are sitting on your couch and think of something that might add value to their project or company, call or e-mail them in the evening and let them know you were thinking about them. Nothing builds your reputation (and stability) better than exemplary customer service skills, especially when your clients are raving to your employer about you. You can't get much more stable than that.

The bottom line is that stability in your engineering career doesn't necessarily come from a paycheck; it comes from you. Rather than defining stability as a regular paycheck and/or benefits, I would define it as your ability to generate regular income and benefits at any time through any situation in your career. The development of your skill sets, your professional network, and your reputation will provide true stability in your engineering career and life.

13.9   Be Cautious, Even When You Find One of the Highest-Paying Engineering Jobs (August 15, 2013)

Many of our IECD members seek coaching when they have found new job opportunities. They seek guidance on whether or not to leave their current jobs and take the leap.

Don't make major engineering career decisions without serious thought and consideration.

The one recurring theme in all of these instances is that the new job offers a higher salary than the old one, which makes the opportunity extremely enticing. Just recently, one of our members told me, "I need to talk. I have an opportunity to take one of the highest-paying engineering jobs in the industry, but I want to discuss it with an engineering career coach first." Good idea, I thought. This post will offer some questions you should ask yourself even when one of the highest-paying engineering jobs is staring you in the face.

Does This Job Give Me a Better Opportunity to Achieve My Goals?

This is the most important question you can ask yourself. As an engineer, you should have clearly defined goals; this is something we instill in our members here at the IECD. Without clearly defined goals, your career may lack direction, and you may lack engagement and motivation. A new career opportunity should always put you in a better position to achieve your goals; if it doesn't, then how much of a career opportunity is it? If one of your goals is to go from engineer to manager and the new opportunity will facilitate that, then I would consider it a good opportunity—unless you are simply chasing an engineering manager salary.

How Can I Minimize Risk When Embracing This New Opportunity?

If you are leaving a stable engineering job for another opportunity, you must consider how you can minimize the risk in your new position. Simply taking one of the highest-paying engineering jobs available is not minimizing risk. There is still risk that it won't work out and in 3 months you will be looking for another job. Be creative and think of ways you can minimize the risk associated with the worst-case scenario.

I spoke to an IECD member recently who has an opportunity to leave a very stable job and start a new office for another company. This is a huge opportunity for him but also a huge risk. We brainstormed ways that he could minimize risk with this new position, and we came up with a plan for him to request that the company give him a one-year agreement minimum, which would give him both stability and time to succeed.

Do I Have a Solid Plan for Making the Most of This Opportunity?

So you found one of the highest-paying engineering jobs that will allow you to achieve your goals faster, but do you have a specific plan to do so? Just because a job, no matter what the salary, offers opportunities for achieving your goals, that doesn't mean you can take it and expect your goals to be realized without you having a solid plan. Before accepting a new position, you have to do your research and create a plan for how you are going to succeed. If you are starting an engineering firm, you'll need a business plan. If you will be working for someone else, you still need a plan with the steps you will take and resources you will need to succeed in that position. Just because the salary is high or the company has a great reputation doesn't mean success is guaranteed in any way.

Interview Yourself

So next time one of the highest-paying engineering jobs or another great engineering opportunity is staring you in the face, be sure to interview yourself and ask yourself these three questions. The results may help you make one of the most important decisions in your engineering career and life.

13.10 If You Set Lofty Goals, You Will Engineer Their Reality (October 22, 2013)

I often talk to our IECD members about setting lofty goals in their engineering careers. These are really big goals—goals that on first thought might even seem unreachable. Once you set these types of goals, it is imperative that you remind yourself of them over and over again, ideally on a daily basis. This can be done through a journal or some other tool capturing your thoughts and goals. I am not just telling IECD members to do this because it is written about in many books; it bears repeating because I have seen it work firsthand in my career and life. In this post, I am going to share three real-life examples from my own experiences.

When you set goals and then keep them fresh in your mind, you start to train yourself both mentally and physically to constantly move toward those goals, whether you realize it or not. At times, actions may even be driven by your subconscious mind because you have trained it as to what direction you want to head in your career and life.

Obtaining My PE License

I knew from the day I graduated college—in fact, I knew before that—that I wanted to get my PE license. Because this was such a clear goal of mine, I was always preparing myself for the exam, even years before I took it. I kept a very good record of all of the projects I worked on from as soon as I started my career. I made sure that I was clear on the kind of engineering work that the state board expected on a PE application, and I made sure that was the work I did.

Early on, I obtained an application from a colleague was previously approved to sit for the exam to ensure I completed mine in the same format. Then, when it came time to study for the PE exam, believe or not, I typed out the words *Anthony Fasano, PE*, and taped them to the top of my computer monitor so that I stared at them all day. Not only did this remind me of my goal, but whenever I tried to come up with a reason not to study, my goal was staring me in the face driving me to do so.

The bottom line is that I took and passed the exam on my first attempt and became one of the youngest people to do so in the State of New York at the age of 24.

Becoming a Partner in My Engineering Company

Another very clear goal that I set for myself early on in my engineering career was to become a partner in my engineering company. Knowing this kept me motivated to develop all of my nontechnical skills, which facilitated my path toward this goal. Also because I was so clear on this goal, I would discuss it frequently with my supervisor and ask him on a regular basis how I was progressing and if there were anything else I should be doing. All of these actions allowed me to become the youngest associate partner in my company, which is a very reputable engineering firm, at the age of 27.

Visiting a Group of Alaskan Huskies

Goal achieved! Ever since I was a kid, I have been fascinated with wolves and wolflike dogs. I am not sure why, but I would always read books about them, watch movies on the subject, and of course hang posters on my wall. I even adopted a wolf once in another part of the country to support the species.

I have always had a goal of either getting a Siberian husky or similar type of dog or spending some time with these animals, but was never exactly sure how or when that would happen. Recently, I was asked to spend a week in Alaska to do a series of my Engineer Your Own Success seminars for local engineers. As soon as I received the call, the first thing I thought about was the Alaskan husky sled dogs and the potential opportunity to see them in Alaska. Because this goal has always been so prevalent in my mind, as soon as the trip went into the planning process, I started researching online how I could visit these dogs. I ended up stumbling upon a civil engineer and his wife who are both dog mushers who have raced in the Iditarod several times and now raise the Alaskan huskies on their property just north of Fairbanks, Alaska. A musher is the driver of a dogsled, and the Iditarod is an annual long-distance sled dog race run in early March from Anchorage to Nome.

I immediately contacted these people, and they invited me to come spend some time with them and their 40 dogs! Sure enough, a week and a half ago, not only did I visit their house in Alaska, but they took me on a 10 mile run with the dogs. Yes, I got to sit on the all-terrain vehicle (ATV) (which they use to pace the dogs) and watch these brilliant animals that I grew up loving do what they do best: RUN. I am still in awe that I was able to do something like that in a place I never even thought I would go during my lifetime. It just goes to show you what your mind will help you accomplish when you are clear on your goals.

13.11 Seven Keys to Success for Engineers and Alaskan Sled Dogs (November 14, 2013)

It's always been a dream of mine to actually see a team of Alaskan sled dogs in person. Last month, not only did I get to realize this dream, but I had the opportunity to actually ride behind these inspiring dogs for a 10 mile scenic ride through the woods of interior Alaska.

In that 30 minute ride, I learned a tremendous amount about these Alaskan sled dogs, which are truly amazing animals. After reflecting on my visit with them, I found seven keys to their success that can be directly translated into an engineering career.

This link will take you to a video of footage from my ride with these dogs, and then below that is my list of seven keys to success for engineers and Alaskan sled dogs based on my experience (http://www.youtube.com/watch?feature=player_embedded&v=8d1-l9BTEWY):

1. **Have goals to strive for**. The sled dogs have a goal of finishing the race to the end. They don't really understand the concept of winning; they just understand that there is an endpoint they are moving toward. Your endpoints in your engineering career are goals—whether finding your first engineering job, obtaining your engineering license, or becoming a partner in one of the best engineering consulting firms. These goals or endpoints give you something to move toward every day.

2. **Pace yourself**. When I rode with the sled dogs, I was sitting on an ATV with their owner. The owner had the ATV in neutral but had it set so the dogs couldn't pull it any faster than eight miles per hour. Why? If we were on a sled, the dogs (especially the younger ones) would start off using as much energy as possible and tire themselves out early on in the run. By running with the 8 miles per hour setting, the dogs learn to pace themselves, so when race day comes, they move at a consistent pace for the entire race. As an engineer, you must pace yourself, too. You should definitely set lofty goals and pursue them, but don't try to do too many things at once or you might burn yourself out.

3. **Know that the path is always changing**. As you will see in the video, the path that the dogs run on constantly changes. They go from an open field to mud to water. Your engineering career path will also change. You may go from engineering school to private industry to a government job and then onto other opportunities. Remember that change is good if you embrace it and find the opportunity in every situation that you encounter. Don't fear change. Embrace it like the sled dogs do on every turn. See the video at http://tinyurl.com/fasano-huskydogs.

4. **Seek motivation from others**. A musher is defined as the driver of the dogsled, but in reality, he or she is much more than that. They train, feed, clean, care for, and love the dogs. When the dogs veer off the path (literally), they are there to refocus them and get them back on track. Don't go at your engineering career alone. It's too difficult to navigate by yourself, and like with anything else, it is always invaluable to have the perspective of others. That is why I created the

IECD to provide career coaching and guidance for engineers. Please take advantage of it. Everyone veers off the trail once in a while, but those who can quickly regroup, refocus, and get back on the track to their goals are the ones who reach the finish line successfully.

5. **Eat right**. As you can imagine, running as much as the sled dogs do requires a lot of energy. Therefore, these dogs' diet is critical to their success and usually includes very high-protein foods. As odd as this sounds, your diet as an engineer is critical to your success as well. The saying "You are what you eat" is worth considering when you plan your meals. To be a top engineer in a top engineering company, you will have to deal with aggressive project deadlines on a regular basis. Eat foods that are high in protein and will give you energy to attack and accomplish your goals.

6. **Be prepared for a long haul**. Alaskan sled dogs are not thrust into big races immediately when their running careers begin. Mushers understand that the dogs need to be trained properly and paced, as most dogs will run many races over their running careers. The same goes for engineers. You will not graduate from college and become a partner in an engineering firm in the same year. Your career is a long haul. Recognize that, and set clear career goals with timelines and small steps toward each one. It is a long haul, but if you take small steps each day, you can achieve your goals faster than you may think.

7. **Get some rest**. While Alaskan sled dogs can run for 20 out of 24 hours a day during big races, they do need those four hours of rest. You, too, as an engineer need rest. There will always be pressure to do better in your engineering career. Produce a better design, bring in more work for your engineering company, get that next degree or certification, etc. Just remember that while it is good to be aggressive, you also need downtime to rest, repair, and reflect. Try to schedule consistent times to do these things over the course of the week, and you will reap the benefits of these actions over the long term.

I thank Judith Currier and her husband Devon (a civil engineer), who allowed me to visit their home for this amazing encounter with these sled dogs. They have both raced the Iditarod several times, and Judy is now training for the Yukon Quest 1000 Mile International Sled Dog Race. I wish her and the dogs the best of luck and wish you the best in your engineering career.

13.12   Do All Engineers Need to Check Things Off to Feel Productive? (December 11, 2013)

I admit it—if I do not have a list of tasks in front of me to check off as I work throughout the day, I am not productive. Is this because I am an engineer? Maybe. However, I can tell you that doing this dramatically increases my productivity level, and in this post, I will share some reasons why. The checklist method, so to speak, has enabled me to get about 12 hours of work done in an eight hour day. I have measured this based on past experience and discussions with other engineers.

Here are five reasons why having all of your tasks laid out in front of you helps you accomplish more:

1. **Facilitates delegation**. When you can view all of your tasks, you can easily identify which of them can be delegated to staff or other team members. This should be done first thing in the morning, so they can accomplish these tasks while you work on others.

2. **Simplifies prioritization**. When you have a dashboard (whether it be online or on paper) that allows you to easily view all of your to-do items, it's easier to prioritize them. This is critical, because if it's not done properly, you might be working on a task that is much less important than three or four other ones on your list, which can really hamper overall productivity.

3. **Eliminates mishaps**. When you work at a fast pace, you have to be sure you don't miss things. Having all of your tasks in front of you ensures that you can see everything and nothing will escape your mind. When you work off the cuff and deal with things as they arise, other things get missed. Missing things can be especially dangerous as an engineer, when your work often has impacts on the health and safety of the public (e.g., roadway and/or bridge design).

4. **Alleviates overwhelming feelings**. For me, when I have a lot of things to do (which is every day), I tend to get flustered or overwhelmed. Usually, after listing and prioritizing all of my tasks, I immediately feel better and get myself into a better state for working and being productive. This certainly doesn't decrease the number of tasks on my plate, but it makes me feel more in control of them, and that matters a lot.

5. **Boosts confidence**. Nothing makes me feel better than physically crossing a task off my list when completed. The feeling of accomplishment inspires me to jump head on into the next task and keep moving full steam ahead. Like the last, this is probably more of an emotional benefit, but I believe it goes a long way in keeping my productivity high.

13.13 How to Not Mess Up Your Annual Review for Engineers (December 24, 2013)

Over the past week, I received at least 10 e-mails or phone calls from IECD members asking if I could help them prepare for their annual performance reviews. Most of the e-mails or phone calls started like this: "Anthony, I have my annual review tomorrow, and I was hoping I could speak to you about it." These calls and requests inspired me to write this post to emphasize the importance of this yearly event and give out some pointers for not messing it up.

Your annual performance review is a great opportunity to reflect upon your career progress and maybe your one opportunity each year to present this progress to your employer. Here are five things you can do to get the most out of your annual review and not mess it up like many engineers do:

Prepare ahead of time. Most engineers, as I indicated previously, wait until a day or two before to start preparing or even thinking about their annual reviews. If you had a big project deadline on Friday, would you wait until Thursday night to work on it? No, of course not. Make your review just as important as all of your work deadlines—otherwise, you are putting your development last, and that is never a good thing. Start preparing one to three months before the actual review.

Spend time reflecting and writing. A few months before your review, reflect on the year that you have had. Make notes of all of your accomplishments as well as things you would like to improve upon. Where possible, quantify your accomplishments using specifics, like this: "Brought in $50,000 worth of business to the firm." Use these notes to either prepare a formal document that you can submit to your supervisor or to answer a prereview questionnaire, if your supervisor gives one to you. This document should be submitted two months before your review to give your supervisor time to review, digest, and utilize the information in evaluating your performance, raise, and potential promotion.

Be clear on your goals. During the actual review (and in your documentation that you submit), you should review all of your goals with your supervisor so he or she understands what you want to accomplish. You should present these goals in a way that reflects the benefits that accomplishing this goal will bring to the company. For example, "I would like to bring in $100,000 worth of new business to help increase our company's profit margin." Do not write, "I would like to bring in $100,000 in new business so that I get promoted to Principal."

Understand what is expected of you. Your review should be a back-and-forth conversation. Just as you should lay out your goals, you should also ask your reviewers what they expect from you over the next year. This is very important but seldom done by engineers. It's great to have goals, but if they are inconsistent with what your company expects of you, the end result will most likely not be good. This is usually the reason engineers don't get promotions they are shooting for. They didn't accomplish what their companies wanted them to. It doesn't mean that they weren't productive or effective, but their productivity wasn't focused where their employers wanted it to be—and this problem could have been avoided through simple conversations during their reviews.

Thank your supervisor. Always thank your supervisor for his or her support over the course of the year during your review. When people feel appreciated, they will go out of their way to provide even more assistance. This should be done whether or not he or she thanks you for your hard work. Do the right thing—it will pay off.

While this post contains only five simple tips, I promise you, if implemented, they can drastically improve your review process and ultimately your whole engineering career and life!

13.14 Three Steps to Becoming a Partner in an Engineering Firm, Directly from an Engineering Partner (February 5, 2014)

I feel lucky that I have the ability to speak to motivated engineers around the world, and I try to use this blog to share valuable information with other engineers who may find it useful.

Recently, one of our IECD members shared with me some information that I found extremely valuable—priceless, in fact.

Most engineers I talk to have the goal of becoming partners in their engineering firms. That's exactly what this member wanted, and she took it one step further. She asked a partner in her engineering firm, point blank, "What do I have to do to be a partner in this firm?" His answer was invaluable. He told her that she needed to do three things, as listed below.

The Triumvirate of Becoming an Engineering Partner

1. **Bring in business.** This is probably an obvious one, and I have written about this before, but it is not easy to do. It requires that you really develop your networking skills and build strong relationships. Most engineers, in fact most professionals, don't know how to do this; it's not easy. The good news is that you can learn—and you will have to if you want to climb the corporate ladder as an engineer.

2. **Be the person the clients want on the job or the reason they retained the firm.** I love this one, and honestly it again ties back to the ability to network, build relationships, and service your clients. Of course, having great technical knowledge will play into this one as well, which is why you should develop both technical and nontechnical skills at all times throughout your career. This is by no means easy, but no one said becoming a partner was easy. This is a powerful skill to have, though. If people are hiring your company because of you, you are a linchpin, my friend. A linchpin is defined in the dictionary as a person or thing vital to an enterprise or organization. This is obviously a pretty good skill or characteristic to possess.

3. **Get along with the other partners.** I can attest to this being 100% true from my days as a design engineer. There are a lot of politics that go into this process, and staying on the good side of the other partners will help you get into their "club." I don't recommend kissing up to anyone or being unnatural to accomplish this, but be yourself and, again, build strong relationships.

Appendix: Recommended Reading

I know…you just read this book and probably don't have time to read many more right now. While this book should have given you the tools that you need to develop your core skills, you may feel as if you need more help on one or more of the skills. To that end, I have gathered some of my favorite resources below. They are categorized by the key elements of an extraordinary engineering career; however, in most cases, the books provide information on more than just that one topic.

For additional book reviews, visit www.EngineeringCareerCoach.com/bookclub.

Career Goals

The Complete Idiot's Guide to Working Less, Earning More by Jeff Cohen

The Complete Idiot's Guide to Working Less, Earning More is for people who are feeling overworked and underpaid. This book is about working fewer hours and increasing your income to ultimately live a better and more comfortable life.

Infinite Possibilities: The Art of Living Your Dreams by Mike Dooley

Infinite Possibilities provides principles that transcend belief, knowing the truth about our human nature and exploring how powerful we truly are. Mile Dooley explains that we create our own reality and our own fate. He inspired me to start writing inspriational messages for subscribers at www.EngineeringCareerCoach.com.

Awaken the Giant Within by Anthony Robbins

This book is about taking control of your life. Anthony Robbins provides some effective strategies in taking charge of your emotions, finances, and relationships. *Awaken the Giant Within* will enable you to discover your true purpose and take control of your destiny. Robbins was the first personal development author that I ever read—his books are very powerful.

Engineer Your Own Success: 7 Key Elements to Creating an Extraordinary Engineering Career,
First Edition. Anthony Fasano.
© 2015 The Institute of Electrical and Electronics Engineers, Inc. Published 2015 by John Wiley & Sons, Inc.

The Magic of Thinking Big by David J. Schwartz

Setting your goals high is what this book is all about. *The Magic of Thinking Big* provides useful methods that you can apply to be successful and motivated. Dr. Schwartz gives strategies for finding success, greater happiness, and peace of mind through thinking big.

Create Your Own Future: How to Master the 12 Critical Factors of Unlimited Success by Brian Tracy

Create Your Own Future is a powerful book that offers 12 principles for success with real-world action plans. I am a huge fan of Brian Tracy because his books are very action-oriented providing specific steps to take to improve your results in life.

Think and Grow Rich by Napoleon Hill

Napoleon Hill has researched over 40 millionaires and found out how they get to be where they are. He provides some clear powerful advice on how to grow rich (not just financially), and reading this book will give you an edge over everyone else if you apply these secrets to success. This is an all-time classic in the personal development genre.

The Alchemist by Paul Coelho

The Alchemist is a magical story about an Andalusian shepherd boy who travels from his home in Spain to the Egyptian desert in search of a worldly treasure. Along his journey, he meets an alchemist and finds the treasures of life like wisdom and following your dreams. This fable will help you to understand the power of following your passion.

Communication

Speak to Win: How to Present with Power in Any Situation by Brian Tracy

Speak to Win is another gem by Brian Tracy in which he reveals some secrets that readers can use to be able to speak with confidence and deliver winning presentations. This ultimate guide can help engineers accelerate their career and achieve even the most impossible-seeming goals—because if you can present clearly, you will control your own destiny.

Networking

How to Win Friends and Influence People by Dale Carnegie

How to Win Friends and Influence People has been helping thousands of people climb the ladder of success in their personal lives and career. This bestseller gives some fundamental techniques in handling people, making people like you, and winning them over with the way you think. This is one of the best self-help books, and it will help you achieve your maximum potential. I listen to Carnegie's advice regularly.

Book Yourself Solid by Michael Port

Michael Port has revealed why self-promotion is a critical factor to success with some guides on how to get more clients and increase profits. This book promotes some key marketing plans and strategies that may be helpful for engineers charged with bringing business into their firm or are building their own business.

Influence: The Psychology of Persuasion by Robert B. Cialdini

Dr. Cialdini has shared his studies on what moves people to change their behavior in this popular book. Reading it will help you to learn six universal principles of influence, how to use them in your career, and personal development efforts.

Never Eat Alone by Keith Ferrazzi

I discussed the power of building genuine relationships throughout this book. In *Never Eat Alone*, Ferrazzi lays out specific steps to doing so and connecting with other people in a way that yields results.

Organization

Getting Things Done: The Art of Stress-Free Productivity by David Allen

Through this book, author David Allen transforms the way you work by providing the secret to stress-free productivity. According to his principle, productivity is directly proportional to our ability to relax. This book has created what is known as the GTD revolution, and it will help you get both mentally and physically organized.

The 4-Hour Workweek: Escape 9-5, Live Anywhere, and Join the New Rich by Timothy Ferris

In this book, Tim Ferris provides a step-by-step guide on how to be successful without working 80 hours per week. This lifestyle design change can transform your life (it did for me) with tips on how to trade a long haul, tiring career into living more, and working less. I owe Ferris a huge "thank you" because this book changed the way I approach my career and life in general—for that I will be forever grateful.

The Relaxation Response by Herbert Benson and Miriam Z. Klipper

The Relaxation Response describes an effective method to relieve stress. The techniques shared by Dr. Benson and his colleagues have been used by health care professionals in treating patients suffering from heart conditions, high blood pressure, insomnia, and other physical problems. Applying some of these techniques in your engineering career could drastically reduce your stress levels.

The Monk Who Sold His Ferrari by Robin Sharma

This top bestseller shows how a lawyer finds wisdom and meaning during his spiritual crisis. This book will not only inspire you but also provide a step-by-step approach to living with a life of passion, purpose, and balance. I regularly review and practice the Ten Rituals of Radiant Living that Sharma covers in detail.

The Power of Less: The Fine Art of Limiting Yourself to the Essential ... in Business and Life by Leo Babauta

By eliminating the unnecessary and identifying the essential, you can focus on accomplishing important goals and be successful in life. In this book, Babauta offers some useful tips on how to break down goals into manageable tasks, create new and productive habits, and increase efficiency. Babauta has successfully provided resources that will help readers to be productive with less. This is a must read if you are feeling stressed and overwhelmed on a daily basis.

The 80/20 Principle: The Secret to Success by Achieving More with Less by Richard Koch

In this book, Koch shares his secret to success using the 80/20 principle. This principle is about being more productive with much less effort, time, and resources. Highly effective people and organizations have employed the 80/20 Principle throughout the world and you will be amazed at how simple this principle is. Don't just read this book; apply the principle in everything you do.

Leadership

Teamwork 101: What Every Leader Needs to Know by John C. Maxwell

Teamwork 101 provides many strategies for developing the leader inside of you. Many of the strategies in this book relate to effective communication and could also have been placed in that category. Maxwell is one of the most accomplished authors on the topic of leadership and has many other wonderful books in this field as well.

Real Power: Business Lessons from the Tao Te Ching by James A. Aurty & Stephen Mitchell

In this book, the authors presented a modern day guide for business leaders based on the world's oldest leadership manual Tao Te Ching from the legendary teacher Lao-Tzu. They claim that the real power lies in the ability to transform a workplace from a source of stress into creativity and joy that eventually leads to business success. While I think it is difficult to gain agreement on some of these strategies in our fast-paced world, I agree they are extremely effective once implemented.

The One Minute Manager by Spencer Johnson and Ken H. Blanchard

This short but interesting book uses a story of a successful manager to present powerful ways to increase productivity, job satisfaction, and personal success that you can apply instantly in your career.

Drive From Within by Michael Jordan

This book is about how Michael Jordan approaches life, sports, and business. He describes how his phenomenal success was attributed to his teachers, mentors, and friends who have guided him throughout his life. What I love about this book is that Jordan's burning desire to succeed jumps off the pages.

Energy Leadership by Bruce D. Schneider

The *Energy Leadership* principle has inspired many managers and leaders (including me) to achieve extraordinary results in whatever they do. This book motivates leaders into reaching their true potential and overcome any obstacles to achieving success. It will challenge you to look closely for the opportunity in every "problem" that you are faced with in your engineering career and life.

About the Author

Anthony Fasano, P.E., LEED AP, ACC

Civil Engineer, Executive Coach, Speaker & Author

Anthony Fasano, AKA Your Professional Partner, CEO, and founder of Powerful Purpose Associates, is a nationally recognized professional coach and inspirational speaker as well as the author of the popular *A Daily Boost from Your Professional Partner*. Fasano has both a B.S. and M.S. in Civil & Environmental Engineering from Lafayette College and Columbia University, respectively.

After spending 10 years as a design engineer, Fasano realized that he had the ability to help engineers and engineering organizations to grow and expand. He now works with engineers through his Institute for Engineering Career Development to help them boost productivity and become successful beyond their expectations while maintaining balance and having fun.

With an appreciative awareness of the stress and challenges of the corporate world, Fasano has dedicated himself to helping engineers in unleashing possibilities to create careers that are exciting, enjoyable, successful, and well balanced.

Anthony Fasano now speaks for organizations around the world on the topics of leadership, business growth, and career development. He also provides group, team, and one-on-one coaching to engineers and engineering organizations. To request a speaking appearance or to learn more about his coaching services, contact him at info@ PowerfulPurpose.com. To receive your free motivational *A Daily Boost from Your Professional Partner*, sign up at www.DailyBoosts.com.

Engineer Your Own Success: 7 Key Elements to Creating an Extraordinary Engineering Career,
First Edition. Anthony Fasano.
© 2015 The Institute of Electrical and Electronics Engineers, Inc. Published 2015 by John Wiley & Sons, Inc.

Index

ABET *see* Accreditation Board for
    Engineering and Technology (ABET)
Accreditation Board for Engineering and
    Technology (ABET), 50
Action exercise worksheet, 37, 141, 155–63
AIAA *see* American Institute of Aeronautics
    and Astronautics (AIAA)
AICHE *see* American Institute of Chemical
    Engineers (AICHE)
AIME *see* American institute of Mining,
    Metallurgical, and Petroleum
    Engineers (AIME)
airplane opportunity, 123
American Institute of Aeronautics and
    Astronautics (AIAA), 93
American Institute of Chemical Engineers
    (AICHE), 93
American institute of Mining, Metallurgical,
    and Petroleum Engineers (AIME), 93
American Society of Civil Engineers (ASCE), 93
American Society of Mechanical Engineers
    (ASME), 93
Annual performance review, 116–17, 189
ASCE *see* American Society of Civil
    Engineers (ASCE)
ASME *see* American Society of Mechanical
    Engineers (ASME)
awards, 58

Babauta, Leo, 108, 118–20, 125, 174, 196
benefits, 58

Benson, Herbert, 136 *see also The Relaxation
    Response*
Blanchard, Ken H., 141, 196
Bluestein, Jeff, 148
Book of Understanding, 132
boss/supervisor, dealing with, 99–101
business, bringing in, 94, 132, 191
business cards, managing, 112–13

calendar, use of, 112–14, 118
career goals, 2, 31, 171, 185
    books, 193
communication, 2, 73–88, 173, 194
    books on, 194
    clear, 77
    early, 76
    framing requests, 79–80
    gaining confidence, 84
    guidelines, 75
    honest, 77–9
    importance of acknowledgement, 85
    mix up, avoiding, 74
    politeness, 79–82
    power of listening, 85, 91
    project/team, 74–5
    public speaking, 80–84
    responsiveness, 86–7
    technical information for nontechnical
        people, 73, 76–7
    tone, 79–82
credentials, 2, 45–7

Engineer Your Own Success: 7 Key Elements to Creating an Extraordinary Engineering Career,
First Edition. Anthony Fasano.
© 2015 The Institute of Electrical and Electronics Engineers, Inc. Published 2015 by John Wiley & Sons, Inc.

Books in the IEEE
PRESS SERIES ON PROFESSIONAL ENGINEERING COMMUNICATION

Sponsored by IEEE Professional Communication Society

Series Editor: Traci Nathans-Kelly

This series from IEEE's Professional Communication Society addresses professional communication elements, techniques, concerns, and issues. Created for engineers, technicians, academic administration/faculty, students, and technical communicators in related industries, this series meets a need for a targeted set of materials that focus on very real, daily, on-site communication needs. Using examples and expertise gleaned from engineers and their colleagues, this series aims to produce practical resources for today's professionals and pre-professionals.

Information Overload: An International Challenge for Professional Engineers and Technical Communicators

    Judith B. Strother, Jan M. Ulijn, Zohra Fazal

Negotiating Cultures: Narrating Intercultural Engineering and Technical Communication

    Han Yu and Gerald Savage

Slide Rules: Design, Build, and Archive Presentations in the Engineering and Technical Fields

    Traci Nathans-Kelly and Christine G. Nicometo

A Scientific Approach to Writing for Engineers & Scientists

    Robert E. Berger

Engineer Your Own Success: 7 Key Elements to Creating an Extraordinary Engineering Career

    Anthony Fasano

3638 030

CPSIA information can be obtained
at www.ICGtesting.com
Printed in the USA
LVOW04s1658040318
568599LV00009B/73/P